The Teaching and Learning of Algorithms in School Mathematics

Lorna J. Morrow
1998 Yearbook Editor
Independent Mathematics Consultant
Aurora, Ontario

Margaret J. Kenney
General Yearbook Editor
Boston College

1998 Yearbook

National Council of Teachers of Mathematics

Library of Congress Cataloging-in-Publication Data:

The teaching and learning of algorithms in school mathematics / Lorna J. Morrow, 1998 yearbook editor, Margaret J. Kenney, general yearbook editor.
p. cm. — (Yearbook ; 1998)
Includes bibliographical references.
ISBN 0-87353-440-9 (hardcover)
1. Algorithms—Study and teaching. I. Morrow, Lorna J. II. Kenney, Margaret J. III. Series: Yearbook (National Council of Teachers of Mathematics) ; 1998.
QA1.N3 1998
[.QA9.58]
510 s—dc21
[511′.8] 98-3128

CIP

Pictured on the cover is a detail of the circuitry on a computer motherboard, the conduits that enable a computer to perform algorithmic functions.

Cover photographs by Jacqueline Marni Olkin

The publications of the National Council of Teachers of Mathematics present a variety of viewpoints. The views expressed or implied in this publication, unless otherwise noted, should not be interpreted as official positions of the Council.

Printed in the United States of America

Contents

Part 2: History

Part 3: Curriculum and Instruction—Elementary Grades

Part 4: Curriculum and Instruction—Middle Grades

Part 5: Curriculum and Instruction—Secondary Grades

Preface

The publication of NCTM's *Curriculum and Evaluation Standards for School Mathematics* in 1989 with its call for decreased attention to be paid to paper-and-pencil computations has provoked many questions about the place of algorithms in today's curricula. This 1998 NCTM Yearbook, *The Teaching and Learning of Algorithms in School Mathematics,* is an attempt to answer many of these questions and to stimulate other questions that all of us in mathematics education need to consider as we continually adapt school mathematics for the twenty-first century.

The papers cover a wide variety of topics, including student-invented algorithms, the assessment of such algorithms, algorithms from history and other cultures, ways that algorithms grow and change, and the importance of algorithms in a technological world. The reader is invited to investigate algorithms from addition and subtraction in the early grades to fractals in secondary or postsecondary studies.

Although the yearbook is divided into five sections, the placement of a paper was not always a simple choice. Because most submissions touched on more than one important aspect of algorithms, papers were placed in sections that seemed to reflect the main issue being discussed.

Part 1, "Issues," treats the nature of algorithms, their importance in a calculator and computer world, suggestions for assessing students' thinking about algorithms, and several definitions of *algorithm* with specific references to particular areas of mathematics. The first paper presents an overview of the issues raised by the call for papers and includes interviews with mathematics educators.

Part 2, "Historical Issues," tries to answer such questions as these:

- Where have our current algorithms come from?
- How were they developed?
- How have they changed?
- What number concepts lie behind early efforts at computation?

Several useful suggestions for projects or history-of-mathematics topics are given in this section while the nature of algorithms is explored.

Many of the papers in Part 3, "Teaching and Learning Algorithms in the Elementary School," discuss student-invented algorithms. Different positions on this important topic are taken. They are presented here as the reported results of many educators' experiences in attempting to empower their students mathematically. One of the most useful aspects of this section is the numerous examples of student thinking about the algorithms that are discussed.

Several papers in Part 4, "Teaching and Learning Algorithms in the Middle School," deal with the bane of middle school teachers: fractions. Accounts of students working together to develop meaningful algorithms show that the ability to communicate mathematically is critical to such an endeavor. The papers also show how an understanding of algorithms can increase a student's data sense or lead students to the generalizability of algorithms and initial experiences with writing algebraic equations.

Perhaps the widest range of topics occurs in Part 5, "Teaching and Learning Algorithms in the Secondary School." Authors discuss ways to help students understand and develop algorithms in graph theory; explore the nature of random numbers through algorithms; use an algorithm to analyze a scatterplot; and apply an algorithm to produce an object that is the basis of a mathematical study.

The creation of a yearbook is a fascinating procedure that brings together many capable and dedicated professionals over the three-year period from developing the call to the final reading of proofs. More than eighty manuscripts were submitted to the Editorial Panel for review. The book is composed of thirty-one of these papers. I would like to thank all those who submitted manuscripts for consideration and in particular those who spent time revising and resubmitting their work. I also want to thank the members of the Editorial Panel, with whom I put in some of the most intense yet enjoyable working sessions of my career, for lending their time and expertise to this project:

George W. Bright	University of North Carolina at Greensboro, Greensboro, North Carolina
John J. Del Grande	Newmarket, Ontario (retired)
Francis (Skip) Fennell	Western Maryland College, Westminster, Maryland
Jacquelyn S. Joyner,	Richmond Public Schools, Richmond, Virginia

Special thanks are due to Margaret Kenney, general editor for the 1996 through 1998 Yearbooks, who was incredibly helpful and patient and always ready with support or guidance. Many thanks, too, go to the editorial and production staff at the NCTM Headquarters Office, who took hundreds of pages of "raw" manuscript and turned them into this finished book, a useful addition to anyone's professional library.

The authors of this yearbook have attempted to answer some of the questions currently facing mathematics educators—and have suggested future directions as well. It is our hope that all teachers will reflect on the ideas presented and use them to adapt their practices to reflect our common goal of empowering our students mathematically.

Lorna J. Morrow
1998 Yearbook Editor

Part 1

1

Whither Algorithms? Mathematics Educators Express Their Views

Lorna J. Morrow

THIS introductory paper takes the form of interviews between the issue editor and selected mathematics educators. The interviews focused on major issues dealing with the teaching and learning of algorithms. Five areas of concern were identified:

1. The value of standard and student-invented algorithms
2. Multicultural issues, such as algorithms from other cultures
3. The role of algorithms in standardized tests
4. The value and place of drill in learning algorithms
5. The place of algorithms in a technological society

Responses to questions on these issues were given by the following:

- Ralph D. Connelly, Brock University, Saint Catharines, Ontario
- Charlie Marion, Lakeland High School, Shrub Oak, New York
- Helene Silverman, Lehman College, New York, New York
- Cathy Seeley, Texas Statewide and Systemic Initiative, Austin, Texas

In each section below, questions about one of the areas of concern are presented, together with responses. Each person is identified by his or her initials. Not all the interviewees were asked to respond to all the questions.

Issue 1: Standard and Student-Developed Algorithms

Many mathematics educators believe that children should be allowed time to develop their own algorithms. Others are convinced, just as strongly, that children should be taught standard algorithms.

- Where do you stand on the teaching of algorithms?

If algorithms are to be taught, decisions need to be made about which algorithms should be taught.

- What algorithms should be considered "standard"? Who should make these decisions (e.g., classroom teachers, mathematics specialists, parents)?

RC: Contrary to what I've heard many math leaders saying, I still believe that there's *some* value in having students learn standard algorithms for computation—after all, these algorithms must have become "standard" for some reason. I think the reason is that they are likely the most efficient. However, we also need to recognize that paper-and-pencil efficiency is now a much less important criterion for the selection of a computational procedure than it has been in the past. In short, I think there's a value in a balance being struck between invented and standard algorithms.

CM: At least some time should be spent each year on activities designed to show that algorithms are the inventions of men and women and not simply discoveries. Teachers should point out that new algorithms are being constructed all the time and show how standard algorithms have changed over the years. At the start of the twentieth century many people were needed who could calculate efficiently and accurately by hand. With the twenty-first century upon us, we need citizens who appreciate how much mathematics contributes to the country's well-being. Which algorithms become standard in the future will depend on the needs of society and the pace of technology. School systems should have some flexibility in deciding which algorithms are taught. Researchers may help decide which algorithms should be taught by showing that little or nothing in the way of understanding is lost when certain algorithms are dropped from the curriculum.

HS: My preservice students find that the questioning needed to help students develop their own algorithms or understand standard algorithms is one of the most difficult skills to learn. Since most teachers teach as they were taught, teachers must first examine their own beliefs about the teaching and learning of algorithms. Learning to value student-invented algorithms is a major change. It is also difficult to envision the end result—that is, what mathematics will the students have learned and how will I, the teacher, know that the student has learned something valuable? Teachers must learn to deal with the insecurity of not knowing exactly what students will be learning. Experience is needed for them to develop these skills.

ISSUE 2: MULTICULTURALISM

In a multicultural classroom students often come to school knowing algorithms they have learned elsewhere.

- Is it important for all students to learn the same algorithm? Why or why not? In what ways, if any, could algorithms from other cultures be used to help students develop a greater understanding of algorithms in general?

RC: I do not believe it's important for all students to learn the same algorithm. Since an algorithm is simply a procedure for arriving at an answer for a particular situation, there would seem to be no reason to insist that everyone use the same one. Indeed, exploring algorithms from other cultures can help students develop number sense by thinking about numbers in different ways.

CM: If children come to school with knowledge of, and proficiency with, an effective algorithm that is different from the standard one for that school, they should certainly be allowed to use that procedure. Furthermore, as long as other students in the class are not expected to learn these different algorithms, some exposure to them may serve to reinforce the message that algorithms are not fixed and universal. Along the way, the resourceful teacher may also use the opportunity to talk about tolerance.

HS: In a city like New York, where students come from all over the world, most classrooms have students who have learned several different algorithms. Those students who have good number sense—especially estimation skills—seem able to make the transfer to a "United States" algorithm. An examination of different algorithms, with the view to understanding them, is a task for both teachers and students. Again, this is an area in which professional development could be of great value.

ISSUE 3: STANDARDIZED TESTS

Standardized tests usually measure a student's ability to use algorithms.

- Is the type of algorithm important?
- Is it possible to change the nature of test questions to encompass a wider variety of algorithms? If so, how; if not, why not? Should the test items be changed? Why or why not?

RC: Most standardized tests with which I'm familiar test a student's ability to get answers to specific questions within a restricted time and usually specify that calculators should not be used. This clearly assumes that the student will use *some* algorithm, but I would argue that it doesn't place any importance on the type of algorithms used. Frequently there's an assumption that standard algorithms will be used, and in the case of multiple-choice responses, the distracters used frequently are based on common errors with standard algorithms. I think it's possible to change the nature of test questions to encompass a wider variety of algorithms—for example, a subtraction question placed in a context in which solving it by adding on makes more sense than using the standard algorithm.

CM: Standardized tests have their place, but with more public reporting of scores than ever, there can't help but be an emphasis on teaching to the test. Consequently, it is extremely important that tests be given that measure the achieving of commonly desired objectives. There's no doubt that standardized testing at the end of a course can result in a restricted learning experience. For example, the test taken by New York State high school math students at the end of their junior year does not allow the use of graphing calculators. Teachers are therefore unlikely to have their students become familiar with such calculators and use them on a regular basis.

HS: A major difficulty with changing standardized tests is that students and teachers need to learn how to deal with open-ended problems. Initial results for such tests are usually very poor. This leaves students feeling less confident about themselves. There is value in assessing reasoning as well as the final answer, but in-service training is needed to help teachers identify what they will be assessing and how the assessment can be done. They also need to learn how to help students value more than just the answer.

ISSUE 4: DRILL

Assuming that algorithms—whether standard or student developed—remain an important part of the curriculum, the question of the rote learning of algorithms must be addressed.

- What should be the role of drill and practice in the learning of algorithms (either student developed or teacher directed)? What is the role of arithmetic games? Of calculators? Of computers?

RC: I still believe it's important for students to practice to develop proficiency in learning and using algorithms. There's also no doubt that students need much less practice to internalize an algorithm than was previously believed. This practice can also take many forms, and much of it can be accomplished through the use of arithmetic games. Using computers only for drill and practice or checking is a gross misuse of the technology. There are computer programs that do a superb job of engaging students in enjoyable practice settings—and even of assisting students to "see" how some algorithms work.

CM: A premium should be placed on studying techniques that can be used not just in mathematics but in any field in which a person needs to learn some procedure or fact so that he or she can perform the procedure or recall the fact effortlessly. For example, why shouldn't students at least hear about the methods actors use to learn many scenes of dialogue? Mnemonics and an emphasis on patterns are two techniques whose application seems universal. Finally, there are times when rote memory must be resorted to. The judicious use of a variety of learning situations should help all students become familiar and comfortable with various algorithms.

ISSUE 5: TECHNOLOGICAL SOCIETY

With the increased use of user-friendly technology, some educators question the amount of class time spent dealing with algorithms at the expense of more-holistic aspects of mathematics such as number sense, spatial sense, problem solving, and data management.

- What should be the place of algorithms in a well-rounded curriculum?

RC: In too many elementary school classrooms, far too much time is still spent on developing proficiency with pencil-and-paper algorithms. This reduces the time available for topics such as number sense. I believe strongly that the more inclusive ideas of number sense—estimation, mental math, a "feel" for large numbers—are more important than algorithmic proficiency and that they need to be given more time and attention in an effective mathematics program. Calculators and computers can offer a variety of contexts for this purpose.

CM: This question is going to be asked over and over again as the pace of technology development increases and cheaper, more-powerful handheld calculators become available to everyone. About the only answers that can be given that will hold true now and in the future are that (1) teachers and students will feel uncomfortable with change, especially when it is dramatic and over a short period of time and (2) research should be initiated that will give a theoretical basis for starting to answer this question and others that arise.

That the importance of some algorithms will change is certain. For example, subjecting all students to the cumbersome steps of the standard algorithm for determining the square root of a number can only serve to convince them that their teacher needs to invest in a five-dollar calculator. What's needed is research into what is gained and lost by working with technology and how teachers can cope with those losses. We should not look at learning a particular algorithm as an all-or-nothing situation. Different levels of mastery of an algorithm may be sufficient for the needs of different students, depending on their interests and talents.

CS: Technology affects mathematics curricula in three ways: First, some topics and concepts will become more important because understanding them is necessary to use technology. Two examples are fraction-decimal equivalents and matrices. Calculators and computers deal, as a whole, with decimals rather than fractions, so students should be conversant with equivalents. Second, other topics will become less important, since the technology can replace them. Many algorithms fall into this class. Educators usually take one of two positions on this issue: One is that the algorithms need not be taught at all, since calculators and computers can do them faster and better. The other is that algorithms should still be taught but the amount of time spent on them should be drastically reduced. Examples are the standard long-division algorithm and the arithmetic of fractions. Third, some topics and concepts become more accessible, and time not spent on algorithms can profitably be spent on some of these topics. For example, problems need not be contrived but can deal with "big, ugly, real numbers."

One other point I would like to make is that with technology we do not have to hold students back if they haven't mastered a particular algorithm. In the past we never knew if these students could deal with the "good stuff" because we did not give them the chance. With the proper use of calculators and computers, these students now have a chance to explore fields of mathematics in which they often show themselves to be proficient. We need to shift our emphasis from a curriculum heavy on algorithms and from the view that all students should become adept at handling algorithms before tackling some of the more interesting aspects of mathematics to a curriculum that explores mathematics in all its wonderful variety.

2

Paper-and-Pencil Algorithms in a Calculator-and-Computer Age

Zalman Usiskin

PAPER-and-pencil work in mathematics begins in kindergarten when children learn how to write numerals. This work proceeds through school with activities such as taking notes, writing the steps of algorithms, recording solutions to problems, and drawing figures. It continues through life, with the recording and processing of information such as that in checkbooks or with the writing of reports. Some of this work involves what we call "skills." The particular paper-and-pencil skills that constitute the focus of this paper are those associated with algorithms. Here we use *algorithm* in a customary way, as a finite, step-by-step procedure for accomplishing a task that we wish to complete (cf. Epp [1990, p. 129]; Lovász [1988]; Musser and Burger [1988, p. 99]).

Types of Algorithms

A great many paper-and-pencil algorithms are used in school mathematics. Sometimes we describe them by the task they accomplish; for instance, we may speak of algorithms for dividing fractions. Sometimes we give them specific names; for example, when we speak of long division, we mean a particular algorithm written down in a particular way. In preparing this paper, I found it impossible to stick exclusively to either of these types of descriptions. But it is possible to sort most algorithms of school mathematics into three categories:

1. *Arithmetic algorithms,* such as those for finding cube roots and square roots; doing long division, long multiplication, column addition, subtraction with many digits, division of fractions, and addition of fractions; and calculating the mean, the standard deviation, and so on
2. *Algebra and calculus algorithms,* such as those for solving linear equations and inequalities, completing the square, manipulating partial fractions, calculating definite integrals, evaluating formulas, using logarithms for computation, simplifying radicals, and determining lines and curves of best fit

3. *Drawing algorithms,* such as those for making bar graphs, circle graphs, coordinate graphs, and graphs of functions and relations; doing ruler-and-compass constructions in geometry; and finding transformation images of figures

This sorting reflects the different kinds of technology that are commonly available as alternatives to paper and pencil. The examples in the remainder of this paper come from these three categories.

BASIC PRINCIPLES ABOUT TEACHING ALGORITHMS

We may not all agree on what constitutes a paper-and-pencil skill or an algorithm. Single-step procedures, such as for the multiplication of fractions, and the procedures implicit in memorized facts are so simple that it is not clear whether we should term them algorithms. However, step-by-step procedures used in certain types of proofs, such as the proofs of formulas for certain sums using mathematical induction or triangle-congruence proofs, are so complex that it is not clear if students are using a prelearned procedure or are solving a new problem when they write an answer.

On some principles, however, virtually all people agree. Some of us might even consider them to be truisms:

1. Technology changes the relative importance of algorithms. Some algorithms become more important, some become less important, and some do not change in importance.

This principle is implicit in the *Curriculum and Evaluation Standards for School Mathematics* (National Council of Teachers of Mathematics [NCTM] 1989, p. 8) and occurs in earlier documents as well.

2. For a given task, algorithms involve three types of processes: those done in the head, those kept track of with paper and pencil, and those done with the aid of technology.

This principle is explicit in the *Standards* (NCTM 1989, p. 94). However, it is not always easy to distinguish among these three types. Consider the multiplication of fractions in algebra—as simple a task as, say, to simplify

$$\frac{3}{x} \cdot x^2.$$

If we write down only the answer $3x$, have we used paper and pencil or done it in our heads? It isn't a basic fact, but we may have done it all mentally, imaging the paper-and-pencil work

$$\frac{3}{x} \cdot x^2 = 3 \cdot \frac{x^2}{x} = 3x.$$

3. No matter what algorithm teachers think they are teaching, students will process it in different ways.

This is a natural conclusion if one accepts the constructivist paradigm that knowledge is constructed by the learner (see, e.g., Noddings [1990] and Goldin [1990]), but one need not be a constructivist to accept its truth.

The construction of knowledge internally does not necessarily imply that there should not be significant external guidance. We each learn language internally but would not learn any English at all if we did not hear or read it. Some of today's algorithms (such as the quadratic formula to solve quadratic equations or the way we multiply whole numbers) have been refined over thousands of years by brilliant people in many different cultures. Thus, except for the simplest tasks, it is expecting too much of students to expect them to invent efficient algorithms. However, it is not only appropriate but advisable to expect students to explore and adapt algorithms.

4. In order to use an algorithm, one must have the necessary tools for that algorithm and must know how to use the tools to carry it out.

The fourth principle applies naturally to calculator and computer algorithms. If we wish students to use technology to carry out a task, we are obligated to teach them how to use the technology. But we cannot expect students to use that technology unless it is readily available to them at all times. This principle explains why the cost of technology plays a significant role in the choice of algorithms.

It also follows from this principle that we cannot expect students to use paper-and-pencil algorithms if they do not have paper and pencils. This conclusion may seem trivial, but in all schools there are students who do not have a sharpened pencil or who have lined notebook paper but do not have plain paper or graph paper when that would be more appropriate. In some urban schools it is not unusual to have a shortage of paper well before the end of the year.

5. To be worth teaching, an algorithm must have a purpose.

Some common purposes of algorithms in school mathematics are to get an answer, to find an equivalent expression, or to draw a picture that makes things easier, clearer, or simpler or that takes less space.

Some standard algorithms have no obvious purpose. For instance, why learn procedures to rewrite

$$\sqrt[4]{1250x^4y^5} \quad \text{as} \quad 5|xy|\sqrt[4]{2y}\ ?$$

Is the latter form easier for later use? Does it help in the understanding of any concept? Does it lead to an easier way to solve any problem? Does it take less space? As another example, why construct the bisector of an angle with ruler and compass? An algorithm is a recipe, and a recipe for something that no one wishes to eat is not particularly useful. If a task has little if any purpose, *any* algorithm for it is not worth teaching.

REASONS FOR ALGORITHMS

If we are going to answer only one question of a particular type, then we do not tend to think of the procedure we used to answer that question as an algorithm. But when there are many questions for which the same procedure works, then it becomes useful to learn the procedure, and once we identify the sequence of steps in the procedure, it becomes an algorithm. Since for most mathematical tasks more than one algorithm exists, then in order to decide which algorithm might be preferred, it is useful to analyze the reasons for having algorithms in the first place.

1. *They are powerful.* Algorithms are generalizations that embody one of the main reasons for studying mathematics—to find ways of solving *classes* of problems. When we know an algorithm, we can complete not just one task but all tasks of a particular kind and we are guaranteed an answer or answers. The power of an algorithm derives from the breadth of its applicability.

The power of some algorithms is easy to determine. A student who has learned a good algorithm for subtracting whole numbers can use this algorithm to subtract whole numbers with any number of digits. The quadratic formula gives solutions for all quadratics, even those with coefficients that are not real numbers. The use of the law of cosines to find the third side of a triangle given two sides and the contained angle works on all triangles.

In this regard, the standard long-division algorithm has more power than using the division key, [÷], on a calculator. Whereas long division can be used with decimals having any number of digits, the use of the division key on a calculator is limited to the number of digits in the divisor, dividend, and quotient that the calculator can hold or display. For instance, it is possible to divide a fifteen-digit number by a three-digit number by using long division with paper and pencil, but most calculators will not allow a fifteen-digit number. Does this mean that we should teach long division? Not necessarily. An algorithm must have a purpose, and if one never has to divide numbers with fifteen digits, then there is no purpose in learning an algorithm for that task.

2. *They are reliable.* The importance of reliability is obvious. When an algorithm is done correctly, it yields the correct answer time after time. When there is a possibility of error in the carrying out of an algorithm, then the algorithm loses some of its utility. So the difficulty of carrying out the steps in an algorithm without an error has to be taken into account when examining it, and the ease with which the algorithm can be reliably applied is one of the reasons that we tend to prefer computer or calculator algorithms to paper-and-pencil ones for some tasks.

3. *They are accurate.* The better algorithms are more accurate, and they may also indicate how accurate the answer is. For these reasons, we are usually not satisfied with an estimate when an algorithm for an exact answer is available.

The Greeks applied this idea to geometric constructions. Because they considered that they were dealing with the ideal world, they were not satisfied with using measurement to find the midpoint of a segment or to build a square whose area was equal to that of a given circle. The square had to be constructed with ideal instruments—the compass and straightedge—not merely measured and drawn. It took two thousand years to determine that the construction of that square was impossible (Beckmann 1982).

Similarly, to determine where a parabola intersects the x-axis, we usually prefer solving a quadratic equation to finding an estimate by examining the graph of the parabola. The SOLVE function on some calculators can do this automatically, but the estimate it gives is not as satisfying as an exact answer. This may also be why we often prefer the exact value 10π to the approximation 31.416; we can find approximations of π to any number of decimal places, so we can approximate 10π as closely as we wish. An algorithm may be preferred when it allows us to determine the result to virtually any prespecified accuracy.

Students have a natural tendency toward accuracy. When a teacher asks a class of students to estimate 51×39 without any context or real-world situation, some will multiply the two numbers, get 1989, and give that as an estimate. The teacher may think they do not understand estimation, but in fact their algorithm (calculate exactly, then give that answer as the estimate) enables them to obtain the closest estimate possible!

4. *They are fast.* An algorithm provides a direct route to the answer and avoids blind alleys, and more-direct algorithms tend to be performed faster than more-circuitous ones. The ease with which the algorithm is learned or recalled is also a factor in the speed with which it can be applied. An algorithm that can be applied fast but that is difficult to remember is not necessarily a good algorithm. For instance, the formula for the inverse of a 2×2 matrix,

$$\text{inverse of} \begin{bmatrix} a & b \\ c & d \end{bmatrix} = \frac{1}{ad - bc} \begin{bmatrix} d & -b \\ -c & a \end{bmatrix},$$

gives an algorithm (substitute) for obtaining that inverse. Since the formula is reasonably simple to use, the algorithm can be performed quickly. But the formula is difficult to recall, and many forget it. An algorithm is less useful and less speedy to use if it is easily forgotten and if one has to find it in a book or derive it each time from scratch.

5. *They furnish a written record.* Paper-and-pencil work, calculator-key sequences, and computer programs can all provide records of the algorithms that were used to determine the answer. The record of an algorithm is significant for teaching because we often want students to examine the process by

which they obtain their answers, to share with one another what they have done, and to refine their procedures. This record also allows us to locate errors in the algorithm more easily. Consequently, an algorithm that operates without a trace, such as often happens when calculators are used, may not be as useful for learning as an algorithm that leaves a trail.

6. *They establish a mental image.* The written record can help establish a mental image that can be used for obtaining results without paper and pencil. For instance, I am always interested in determining the fuel efficiency of my car, that is, the number of miles per gallon it gets. Suppose I drive 412 miles and use 15.4 gallons. To divide 412 by 15.4 in my head, I mentally picture the long-division algorithm, first try 2×154 and place 308 under the 412, record 2, and subtract. Now I image 104 in my mind, divide 1040 by 154 in my head, and obtain 6. This gets me about 26 miles per gallon, and in fact I know that the remainder is about two-thirds of the divisor, so I estimate 26.6 miles per gallon. I use this imaging method with algebra, also. If I wish to solve $8x - 14 = 6x + 88$, I subtract $6x$ from each side and then I add 14 to each side. I get $2x = 102$, from which $x = 51$. My mental methods are clearly derived from the paper-and-pencil methods I was taught.

Students who are taught to obtain quotients by using calculators cannot have a mental picture of long division, so it is possible that students of the future will not be good at complicated mental division. But this would not be much of a loss; most adults today do not image long division as I do even though they may have spent years learning it, and most require paper and pencil to do even the simplest algebra.

7. *They are instructive.* Some algorithms give insight into the relationship between the answer and the given information. For instance, the algorithm used for adding columns of numbers, in which one records a "carry" digit somewhere, is instructive in that it applies the ideas of place value. The multiplication of polynomials in algebra is also instructive—we can identify a clear use of the distributive property when, for instance, we show the middle step in

$$(a - b)(a + b) = a^2 - ba + ab - b^2 = a^2 - b^2.$$

Another way of saying this is that these paper-and-pencil algorithms provide not only a method for getting a result but also a proof of the result or a sketch of a proof. They don't just announce the result; they construct it. As another example, consider the sentence $30 + 0.41m > 50$, which could arise if one were given two choices for renting a car, either \$30 a day and 41¢ a mile or \$50 a day with unlimited mileage, and wondered when the first choice would become more expensive. Here m is a number of miles that for the purposes of computation of cost is rounded up to the nearest integer. How we record the algorithm is important. Here are two recordings:

$30+0.41m>50$	$30+0.41m>50$
	$30+0.41m-30>50-30$
$0.41m>20$	$0.41m>20$
	$\frac{0.41m}{0.41}>\frac{20}{0.41}$
$m>\frac{20}{0.41}$	$m>\frac{20}{0.41}$
$m>48.78\ldots$	$m>48.78\ldots$
$m>48$	$m>48$

Each algorithm proves that the first choice is more expensive when more than 48 miles are driven. Inserting additional steps makes the movement from one step to the next rather obvious, but these extra steps disconnect the given from the solution. By the time we go through all these steps, the student may have forgotten what the question was about. As teachers, we know we need sometimes to put in more steps and sometimes to take out some steps. When the steps are easy, then they are easy to put in or take out; when the steps in an algorithm are tedious, we may wish to omit them always or almost always.

We see this idea of algorithm and proof also in geometry, with a ruler-and-compass construction of an equilateral triangle (see fig. 2.1). This construction is instructive; it elegantly exhibits not only the equilateral triangle but also the characteristics that make it equilateral. However, the ruler-and-compass construction of an angle bisector in figure 2.2 is not very instructive. It does not so closely resemble a proof. Simply measuring half the angle and drawing an appropriate ray is more instructive.

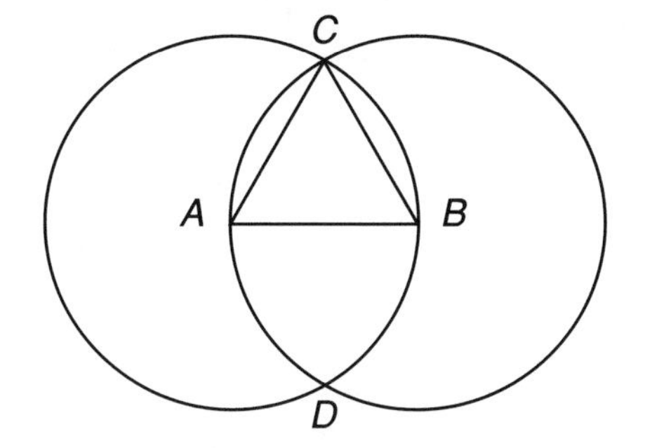

1. Draw circle A, radius AB.
2. Draw circle B, radius AB.
3. Circle $A \cap$ Circle $B = \{C, D\}$.
4. Draw $\overline{AC}$, $\overline{BC}$.

Fig. 2.1. An algorithm for constructing an equilateral triangle with side $\overline{AB}$

Calculator algorithms also have their instructional advantages. Although algorithms on calculators may bypass some intermediate steps, because of their speed they can show patterns between givens and answers that are unattainable or hard to see when using paper and pencil. For instance, we can change factors and see changes in products; we can change coefficients

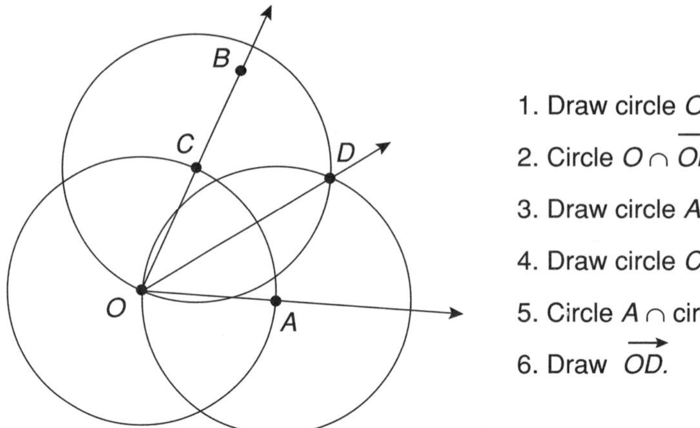

Fig. 2.2. An algorithm for constructing the bisector of $\angle AOB$

of equations and see the effects on the solutions; we can change coefficients of formulas and see the changes in the graphs; we can transform geometric figures and compare the images with one another or with the given figure.

8. *They can be used in other algorithms.* An algorithm used in one situation may be part of a larger, multistep algorithm applied in another situation. For this reason, certain algorithms can be more important than others. Teaching long division is often justified because one algorithm for the division of polynomials follows the same form. The addition and subtraction of positive and negative numbers are needed for solving linear equations. The drawing of a parallelogram is needed for the drawing of a cube (if the cube is not drawn in perspective). The graphing and tracing of a function are needed if we wish to use the graph to estimate a maximum point or a solution to an equation.

9. *They can be objects of study.* We can compare their efficiencies, their mathematical characteristics, and the speed with which they can be performed. For instance, we might compare the algorithms for two different ways to divide fractions:

The procedure on the left requires less writing, but some might believe the procedure on the right is more instructive and more easily remembered.

$$\frac{a}{b} \div \frac{c}{d} = \frac{a}{b} \cdot \frac{d}{c} = \frac{ad}{bc} \qquad \frac{a}{b} \div \frac{c}{d} = \frac{\frac{a}{b}}{\frac{c}{d}} = \frac{\frac{a}{b} \cdot bd}{\frac{c}{d} \cdot bd} = \frac{ad}{bc}$$

Comparisons of the properties of algorithms are commonly done in college-level mathematics courses offered in computer science departments.

DANGERS INHERENT IN ALL ALGORITHMS

Some of the very properties that make algorithms important—speed, reliability, and instructiveness and the mental images they may generate—may create dangers.

1. *Blind acceptance of results.* Because algorithms are reliable if done correctly, answers are often blindly accepted. This is true regardless of whether one does the process mentally, with paper and pencil, or with technology. For this reason, as often as is practical, students should have checks or a second way to get an answer to a question, whether the algorithm is long or short and whether it is easy or hard to apply.

2. *Overzealous application.* Because the best algorithms are powerful in the sense that they can be widely applied, there has always been the tendency for students to overapply them or to use paper and pencil for a task that could be done mentally, as in the following four examples:

A.

$$\begin{array}{r} 345 \\ \times\ 10000 \\ \hline 000 \\ 000 \\ 000 \\ 000 \\ 345 \\ \hline 3450000 \end{array}$$

B.

$$\begin{array}{r} 18.5 \\ 2\overline{)37.0} \\ \underline{2} \\ 17 \\ \underline{16} \\ 10 \\ \underline{10} \end{array}$$

C.

$$8 \times \frac{3}{4} = \frac{8}{1} \times \frac{3}{4} = \frac{2}{1} \times \frac{3}{1} = \frac{6}{1}$$

(The student may not realize that the answer equals 6.)

D.

$$4m = 32$$
$$\frac{4m}{4} = \frac{32}{4}$$
$$m = 8$$

And we all see students use calculators to do arithmetic that they should do in their heads.

These overzealous applications of algorithms may be unwise and inefficient, but they are not wrong. The students are playing safe, worried about losing accuracy if they deviate from an algorithm for a simple special case. In this regard, a common story told to me by adults (who are trying to make the point that students today are not as capable as they used to be) is that they are astonished that cashiers can no longer give change without having the cash register calculate it for them. I have never seen a study of this practice and wonder how widespread it is, but I would rather the cashier have an automatic way to give me change; I trust the machine more than I trust the cashier. And both the cashier and the cashier's boss are probably correct to require that the machine be used whenever possible.

Overapplying mental mathematics can be as dangerous as overusing paper and pencil or a calculator. With mental arithmetic, we have no record of the input, so if there is an error, we often cannot tell whether it is in carrying out the algorithm or in using a mistaken input.

3. *A belief that algorithms train the mind.* Some claim that doing paper-and-pencil algorithms trains the mind in good ways, that going through these

procedures teaches orderliness and care. There has never been any evidence to support this claim. Mathematicians, who tend to be better at these algorithms than other people, seem to be no more orderly.

Algorithms do provide mental images and can be instructive, but these mental images are intended to be applied to a particular task, and there is no evidence that these images generalize to broad abilities such as spatial perception or problem solving.

The evidence is, in fact, to the contrary. Difficult algorithms seem to take students' minds off the bigger picture, keep more important mathematics from being taught, and turn many students away from mathematics.

4. *Helplessness if the technology for the algorithm is not available.* Paper-and-pencil algorithms and calculator or computer algorithms both require the user to have some equipment. Proponents of paper-and-pencil algorithms have often argued that paper and pencils are more likely to be available than calculators or computers, but today the reverse is sometimes true: when schools are out of paper but computers are available; when all the pencil points are broken and the sharpener is out of commission but a calculator is available; or when one needs quick computations and has a tiny calculator in a wallet, pocket, or purse but has no pencil and paper.

Choosing between Paper-and-Pencil and Calculator or Computer Algorithms

When there is more than one algorithm with sufficient power to complete a particular task, the algorithm of choice is normally the faster, the more accurate, or the more reliable. On these dimensions, many paper-and-pencil algorithms are undeniably poor compared with calculator or computer algorithms, which may account for the use of technology rather than paper and pencil by adults as consumers, at play, and in the workplace. The speed of calculators and computers also frees up time in the classroom to study the other dimensions of mathematical understanding—properties, uses, and representations—and to discuss many aspects of problem solving—the relationships of problems to one another and the various ways they can be solved. The speed, accuracy, and reliability allow students to see patterns between problems and solutions that they might not normally be able to see and allow them to explore these relationships. Furthermore, because technology often allows us to go directly from given to result, the intermediate steps do not get in the way. For instance, we do not have to spend time plotting points to see how the equations and graphs of functions might be related.

Here I offer examples of decisions I have made about which algorithm to use for accomplishing a particular task. I am simplistic and compare paper-and-pencil to calculator or computer algorithms as if only one algorithm of each type exists. There are then only four possibilities, as shown in the chart in figure 2.3, where Y stands for Yes and N for No.

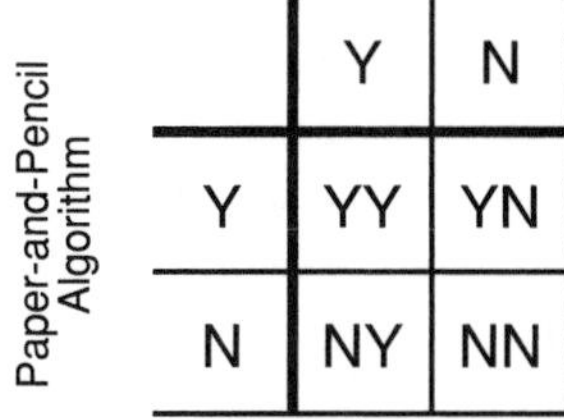

Fig. 2.3

YY: *Teach both paper-and-pencil and calculator or computer algorithms.* The safest decision is to try to have students learn both types of algorithms. Consider operations with fractions. Because fractions represent division in standard algebraic notation, algorithms for operations with arithmetic fractions are applicable also to algebraic fractions. They are more important than algorithms for operations with decimals. They are used, for example, in the solving of equations such as $(2/3)y = 10$ or the rewriting of $3x + 4y = 12$ in slope-intercept form. Consequently, a decision not to give students paper-and-pencil work with fractions may result in these students' having greater difficulty later when using paper-and-pencil algorithms with algebraic fractions. For the same reason, we decide to teach the paper-and-pencil algorithms even when our students have calculators that operate on fractions.

I believe both kinds of algorithms should also be taught for some graphing. The paper-and-pencil process of graphing a line is quite instructive and also accurate, and the need to draw a scale in this first experience of graphing an equation is important to emphasize. Furthermore, to some extent the hand graphing of all equations and functions is more or less the same, and so this paper-and-pencil algorithm does not have to be taught again. But after students learn to graph lines, they should have automatic graphers available because of the speed, accuracy, and reliability of graphing technology.

I also believe that students should be able to draw both with paper and pencil and with the aid of technology because the technology is not always available and can take longer than using simpler tools.

YN: *Teach only paper-and-pencil algorithms.* Every paper-and-pencil algorithm can be programmed to be done by a computer. So when we teach only paper-and-pencil algorithms, it is never because we could not perform the task with technology but because the newer technology is so difficult to use or so costly to purchase that we cannot assume its use.

Software such as Mathematica, Derive, and Maple and the TI-92 and HP calculators do symbol manipulation. This technology is reliable, accurate, fast, and powerful but not always easy to use. The technology can be instructed to go through steps, so it can provide a record as thorough as any paper-and-pencil record. But a record of those steps is not easy to obtain and the technology is not yet cheap enough, and at this time even zealous advocates of the technology have recommended that we not require knowledge of these calculator or computer algorithms in early high school.

However, when symbol manipulators become widely available, we will probably take the same view with equation solving that we do with graphing.

That is, we will continue to teach students paper-and-pencil means for solving linear equations because the idea is important and the process is generalizable, but we will also teach how to use symbol manipulators to solve these and more-complicated equations. The algorithms for equation solving and complicated algebraic manipulation will then move in the table from the right-hand column to the left-hand column.

NY: *Teach only calculator or computer algorithms.* Sometimes we have ignored paper-and-pencil algorithms, even when they were not difficult. Time spent on remedial arithmetic at the seventh- and eighth-grade levels tends merely to put students further behind rather than get them ahead. Thus, for students who perform at or above the seventh-grade level, I believe we should allow the use of calculators and not *reteach* any paper-and-pencil algorithms for decimal arithmetic, even those we consider important. I want students to be able to add and subtract decimals by hand, but they should have learned to do so before seventh grade.

Every calculator or computer program could also be done by paper and pencil, but some paper-and-pencil methods are so unwieldy that they are neither fast nor accurate. Then we may teach only the calculator or computer algorithms. For instance, students should be taught how to use the graphing capability on a graphing calculator to estimate the solutions to numerical equations in one variable. These calculator algorithms are reliable and fast, provide a mental image, and are instructive and so powerful that they can be employed on fifth-degree polynomial equations, or $\sin x = \log 3x - 5$, or many others for which there is no known paper-and-pencil algorithm that gives an exact solution. The only drawback of calculator algorithms is that they do not always give exact solutions. But even this drawback can be turned into an advantage because it motivates students to ask when we can get exact solutions and to examine such equations more closely.

Deleting certain paper-and-pencil algorithms has not handicapped students. Absent from most of today's books are paper-and-pencil algorithms such as Cramer's rule for solving systems; the hand calculation of square roots; the use of logarithms for computation with trigonometric functions or with complicated arithmetic; the drawing of a triangle given the lengths of some combination of sides, medians, or altitudes; and complex manipulations with radicals and with rational expressions. Yet today's students can use technology to accomplish most of the tasks for which these algorithms were designed.

For graphing complicated functions, the accuracy, speed, and reliability of automatic graphers greatly exceed those of paper-and-pencil methods. In fact, the best paper-and-pencil methods for many functions require calculus and so are out of the reach of precalculus students. So we should assume that students have automatic graphers always available and not use paper-and-pencil methods for complicated functions.

For obtaining the line or other curve of best fit, for estimating nth roots or calculating nth powers, or for doing all sorts of complicated arithmetic, it is appropriate to ignore the paper-and-pencil methods and to rely instead on the more accurate, faster, more reliable, and more powerful calculator algorithms.

NN: *Teach neither paper-and-pencil nor calculator or computer algorithms.* When the task is not important enough, then it does not belong in the curriculum. This is obvious enough, but importance itself changes: a task that is not important at one level of schooling may be important at some other level. For instance, operations with rational polynomial expressions are not important enough to be taught in the first two years of algebra because the instructional time should be devoted to more-important things that all students should know. But for a variety of reasons, calculus requires that students be familiar with rational functions, and in a precalculus course, where these functions are discussed, it is suitable to discuss the corresponding manipulations.

PEERING INTO THE FUTURE

We are in a time of change, in which almost year by year teachers tend toward using a calculator or computer (a shift to the left on the chart) and—perhaps less dramatic—toward not teaching as many paper-and-pencil algorithms (a shift down). It is almost certain that we will always teach some algorithms with both paper and pencil and calculators or computers (upper left in the chart).

It seems quite likely that these trends will continue. When the paper-and-pencil algorithms used today were introduced—most about four or five hundred years ago—they constituted the "advanced technology" of their time. Paper was not as widely available as it is today, and the writing implements of the time were much more difficult to use than today's pens and pencils. These algorithms were not immediately accepted by all because some people feared that a loss of mental power would result from algorithms that enabled one to do mathematics with less mental arithmetic and other mental work. Even though paper-and-pencil algorithms ultimately triumphed and became used for almost all mathematics, still one was expected to memorize certain facts and formulas. The old ways did not die out completely, because it was not efficient to have to recall or reconstruct or even write mathematics every time it was needed. And so we can expect that although calculator and computer algorithms will overtake virtually all paper-and-pencil algorithms, a few of these old algorithms will remain. Those that remain will be in our curriculum not because they are curiosities and not because they train the mind but because they have the qualities of good algorithms.

REFERENCES

Beckmann, Petr. *A History of* π. Boulder, Colo.: Golem Press, 1982.

Epp, Susanna. *Discrete Mathematics with Applications.* Belmont, Calif.: Wadsworth, 1990.

Goldin, Gerald A. "Epistemology, Constructivism, and Discovery Learning in Mathematics." In *Constructivist Views on the Teaching and Learning of Mathematics, Journal for Research in Mathematics Education* Monograph No. 4, edited by Robert B. Davis, Carolyn A. Maher, and Nel Noddings, pp. 31–47. Reston, Va.: National Council of Teachers of Mathematics, 1990.

Lovász, László. "Algorithmic Mathematics: An Old Aspect with a New Emphasis." In *Proceedings of the Sixth International Congress on Mathematical Education,* edited by Keith Hirst and Ann Hirst. Budapest: Janos Bolyai Mathematical Society, 1988.

Musser, Gary L., and William F. Burger. *Mathematics for Elementary Teachers.* New York: Macmillan, 1988.

National Council of Teachers of Mathematics. *Curriculum and Evaluation Standards for School Mathematics.* Reston, Va.: National Council of Teachers of Mathematics, 1989.

Noddings, Nel. "Constructivism in Mathematics Education." In *Constructivist Views on the Teaching and Learning of Mathematics, Journal for Research in Mathematics Education* Monograph No. 4, edited by Robert B. Davis, Carolyn A. Maher, and Nel Noddings, pp. 7–18. Reston, Va.: National Council of Teachers of Mathematics, 1990.

3

What Is an Algorithm? What Is an Answer?

Stephen B. Maurer

TODAY, with calculators and computers, the solution to a mathematical problem is often algorithmic. The process of solution, and sometimes even what counts as an answer, differ from before. This paper illustrates the differences with a simple case study—a problem that came up in my own work that I solved with an algorithm. I show how I started, how I reflected on what I did, and how my idea of a solution changed. First, however, I explain what is meant by an algorithm and by modern algorithmic thinking, or algorithmics (Maurer 1992).

WHAT IS AN ALGORITHM?

An algorithm is a precise, systematic method for solving a class of problems. An algorithm takes *input,* follows a *determinate* set of rules, and in a *finite* number of steps gives *output* that provides a *conclusive* answer. *Determinate* means that for each allowed input, the first action is precisely specified and, more generally, that after each action in the sequence the next action is precisely specified. *Conclusive* usually means that the output correctly solves the problem, but it can mean that the algorithm either solves the problem correctly or announces that it cannot solve it.

Example 1. Multiplication

The traditional paper-and-pencil algorithm for multiplying two numbers expressed in Indo-Arabic numerals is brilliant. It is too bad we take it for granted. The algorithm is brilliant because it reduces a general problem to a small subcase—how to multiply two single-digit integers—and does so in a small amount of space. Here is the result of applying the algorithm to 432×378:

```
   432
   378
  ----
  3456
 3024
1296
------
163296
```

Each row of intermediate calculation is obtained by multiplying the top factor (432) by one digit of the bottom factor—a many-digits-times-one-digit subprocedure. To see the basic one-digit-by-one-digit steps, expand the first intermediate row to show more detail:

```
 432
   8
 ---
  16
 24
32
----
3456
```

Of course, the calculation is not usually written this way. To save space, the "carries" are all either done mentally or marked with small digits.

This method is indeed an algorithm: it is determinate, finite, and conclusive. To demonstrate these characteristics formally would take some work. To begin with, we would have to write down the steps of the algorithm carefully, say, using algorithmic language, as in the next example. We never do so with children because a formal description is quite complicated.

Notice that this algorithm involves *iteration* (loops); that is, some subprocess is applied repeatedly. In this instance the subprocess is to multiply two single-digit numbers and carry.

Example 2. The Quadratic Formula

The traditional formula for solving $ax^2 + bx + c = 0$ seems simple enough; where is the algorithm and why bother with it? In example 1 we needed an algorithm because the "formula" for multiplication, $a \times b$, is too simple; it doesn't tell us how to do the multiplication. But for quadratics, if we already know the four basic arithmetic operations, the quadratic formula seems to tell us everything we have to do.

Well, not quite. There are several cases—two distinct real roots, one repeated real root, and no real roots. When there is just one root, we probably don't want to output it twice. When there are complex roots and we are using a calculator that balks at negative square roots, we must compensate. So we can use the following algorithm:

Input: a, b, c (coefficients of $ax^2 + bx + c$, with $a \neq 0$)
Algorithm
  Let $D = b^2 - 4ac$

```
If                                        (three cases follow)
D > 0  then                               (two real roots)
   let s = √D
   let x1 = (−b + s)/2a
   let x2 = (−b − s)/2a
else if D = 0  then let x = −b/2a         (one repeated real root)
else if D < 0  then                       (two complex roots)
   let s = √−D
   let x1 = (−b + is)/2a
   let x2 = (−b − is)/2a
endif
Output: the roots x1 and x2 or the root x
```

This procedure is an algorithm: it is determinate, finite, and conclusive. Although it is longer than the quadratic formula, we would argue that it yields a better answer.

Notice the algorithmic notation. To express an algorithm precisely, one needs a special language. Computer scientists call such notation *pseudocode* because it does not run on any actual computer. I prefer to accent the positive: it is a good notation for human communication about algorithms. Hence I call it *algorithmic language.*

Notice that our algorithm has several if-statements (and no loops). If we want to be even more comprehensive and allow input with $a = 0$, then we have to include several more cases and thus several more if-statements. Procedures in the everyday world—calculating taxes, for instance—often involve many cases. Thus algorithms for everyday problems often include many if-statements.

Example 3. Sequences of Heads and Tails

An important role of mathematics is to guide us in making decisions under uncertainty. We can do so using probability theory, but often the simplest approach is simulation. Suppose we flip a fair coin until we get two heads in a row. How many flips should this take? If we actually carry out this experiment many times, we find out what to expect. The following is an algorithm to carry out the experiment one time. Running it on a computer is faster than flipping coins, though with a whole class we could do the latter (parallel processing!). Rand(0,1) is a command for flipping a coin; the output 1 means heads, 0 means tails. The algorithm could be run a thousand times inside a loop of a bigger algorithm, which could then analyze the output data in various ways (e.g., determine the average and the variance, draw graphs, etc.).

```
Input: (none)
Algorithm
   Let count = headcount = 0
```

```
  Repeat
    count = count + 1
    let flip = Rand(0,1)                 (0 or 1, at random)
    if flip = 1  then let headcount
      = headcount + 1
        else let headcount = 0
  endrepeat when headcount = 2
Output: count                            (total flips to get two heads
                                          in a row)
```

This algorithm violates our definition in two ways. First, when the command generates a random number, it cannot be said to be precisely specified. Second, it is theoretically possible that this procedure will not terminate; we might get 0s forever. Nonetheless, we certainly want to study such "algorithms," and so it is traditional to relax the three defining conditions. The hard part, actually, is to get computers to perform such procedures, since computers really are determinate machines. In other words, how can computers be made to produce what appear to be random numbers? Fortunately, there are good answers, which we will not pursue here.

ALGORITHMICS

The phrase *algorithmic mathematics* has two meanings, traditional and contemporary (Maurer 1984). The traditional meaning emphasizes carrying out algorithms; the contemporary meaning emphasizes developing them, understanding them, and choosing intelligently among different algorithms for the same task.

This contemporary meaning has its own name: *algorithmics*. At the advanced level algorithmics is quite sophisticated. Developing algorithms expands to whole courses in algorithm design (bottom up, top down, backtrack, recursive, etc.). Understanding becomes algorithm validation (mathematical induction is the basic tool). Choosing intelligently becomes complexity theory (precise ways to count or bound the number of steps in an algorithm or needed by any algorithm for a specified problem). In schools algorithmics can be more informal.

In any event, algorithmics does *not* mean performing algorithms over and over by hand. Algorithms will be carried out more and more by machines or by person-machine combinations. Algorithmics is thinking *about* algorithms, not thinking *like* algorithms. For more on the difference between modern algorithmic thinking and traditional problem solving, see, for instance, Hart (1991), Maurer (1984, 1992), Maurer and Ralston (1991), and NCTM (1989, 1990).

THE CASE STUDY

While preparing a contest book (Berzsenyi and Maurer 1997), I needed to create figures to illustrate plane geometry constructions. We were producing the book with the mathematics typesetting system TEX. For figures we were using a macro package called PICTEX; this package has considerable production advantages for use with TEX, but some of the PICTEX commands for geometric constructions are quite primitive.

Consider figure 3.1. In this triangle, *D* is the midpoint of *BC* and *E* is the trisection point of *AC* closer to *A*. With PICTEX it is easy to get all the line segments shown. Just name the coordinates of the endpoints (in some units, say, 0.1 cm). For instance, I declared $A = (0, 0)$, $B = (100, 0)$, and $C = (60, 90)$, in which case $D = (80, 45)$ and $E = (20, 30)$. However, suppose I also needed to mark the intersection point *F*. Of course, PICTEX draws the intersection as it draws the line segments, but it doesn't automatically put down a large dot and a letter. To do so, PICTEX needs to know the coordinates of the intersection point, and it does not compute them itself. So a problem I faced over and over was to find the coordinates of the intersection of two lines when each line was given by two points (see note 1 in appendix 2).

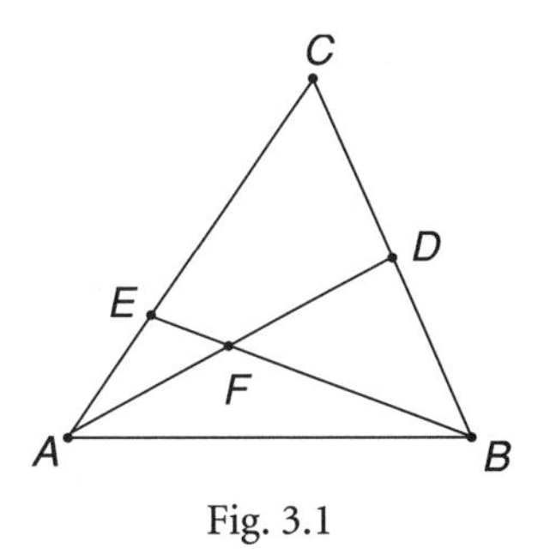

Fig. 3.1

Figure 3.2 is a more complicated example, directly from the contest book. Circle *O* is inscribed in equilateral triangle *ABC*, and *DE* is tangent to the circle at *T* and perpendicular to side *AC* at *E*. I needed to find the coordinates of every labeled point in order to instruct PICTEX to draw the lines and insert the labels. To begin with, I knew the coordinates of *A*, *B*, and *C* because I had created the triangle by choosing them. To illustrate how I found the other points, consider *E*. It is the intersection of the lines *AC* and *DE*. I knew two points on *AC*, namely, *A* and *C*. As for line *DE*, I knew how to find the coordinates of *T* (see appendix 1). From this information it was easy to name another point on *DE*; use the fact that the slope of *DE* is the negative reciprocal of the known slope of *AC*, and go out from *T* along this slope by, say, one unit. Sure enough, I had reduced finding *E* to the problem of finding the intersection point of two lines when each line was defined by two points.

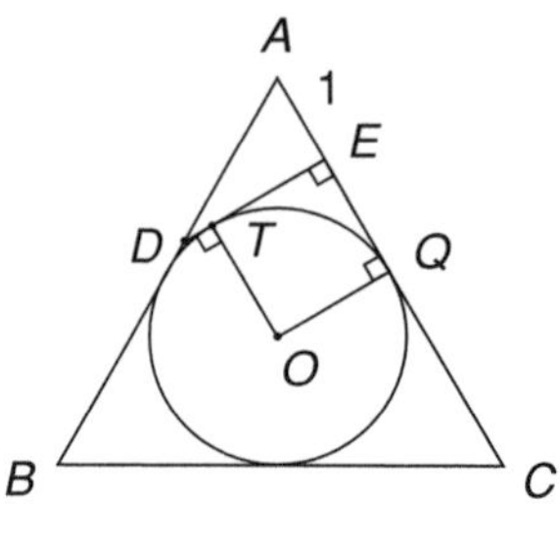

Fig. 3.2

After solving this type of problem by hand once or twice (tedious!), I decided I really needed a little program to do it. I could run this program and edit the book on the same personal computer and paste the program's results into PICTEX.

I planned to write a three-line program in BASIC. The first line would ask for the eight coordinates of the two pairs of defining points as input and assign them to variables. The next two lines would print the value of the formulas for the *x*- and *y*-coordinates of the intersection, formulas expressed in terms of the input variables. I would merely have to derive these formulas once, algebraically by hand, and be done with it.

How would I derive these formulas? Finding the intersection of two lines is a standard problem of algebra 1 if the equations of the lines are given. Using "reduce to the previous case" (a basic strategy of algorithmics), I planned as follows:

1. Use the two defining points for each straight line to find an equation in the form $y = mx + b$.
2. Solve these two equations for x by setting $m_1x + b_1 = m_2x + b_2$.
3. Plug the x solution into $y = m_1x + b_1$ to get the y solution.

All I had to do was substitute the results of each step into the next step to get explicit formulas in terms of the initial inputs. I got out a piece of paper and started.

Well, it was easy in principle but not in fact. The expressions for m_1 and m_2 were modest, but b_1 and b_2 were messier. The x formula was even messier. I didn't want to do step 3.

I could have done it. But I would probably have had to check it three times to get it right. And then I would still probably have made a mistake when I tried to type it into the computer in a form acceptable to the software. I don't have as much patience for such calculations as I did in my youth, and most youths never have much patience for them. Was there a better way with a modern tool?

For a brief moment I considered doing the whole thing in Mathematica. It could have solved all the algebraic equations. But first I would have had to enter the problem with the correct, complicated syntax. Then I would also have had to use the algebraic solution to create a function in Mathematica to compute the numerical values in each particular case because the algebraic output of Mathematica could not go directly into BASIC. And Mathematica is a big program, slow to start up. Using the program did not seem worth it.

So I thought a little harder, and then—I had to laugh at myself. I had not been following the algorithmicist's advice I had given to others so many times: *Formulas are not so important anymore.* Why bother substituting to get one final algebraic formula when I was after numerical answers anyway? I had already solved the problem; I could just leave the results unsubstituted. The BASIC program would have more lines, but it would be much easier to write.

To see what I mean, look at the final True BASIC program I wrote, Algorithm 1. Variable m_1 is the slope of the first straight line; m_2 is the slope of the second. The equations of the two lines are $y = m_1x + k_1$ and $y = m_2x + k_2$. (I used k because I had already used b for the vertical coordinates of the first two points.) I computed k_1 by substituting

$$m_1 = \frac{y - b_1}{x - a_1}$$

into the line equation and rewriting on paper. Next, I set $m_1x + k_1 = m_2x + k_2$ and solved for x on paper; the result is the "let x" command. Then y is obtained directly from the formula for the first straight line. The last little touch concerns the output. I had True BASIC produce exactly the expression I had to plug into PICTEX, not just the proper coordinates. This way, as I edited the book, whenever I needed this program, I just switched to BASIC (without closing the editing program), ran this algorithm, and cut and pasted the output back into my editing file (see note 2 in appendix 2).

Algorithm 1

```
do
    Input prompt "first pair of points ": a1,b1,  a2,b2
    Input prompt "second pair ": c1,d1,c2,d2
    let m1 = (b2–b1)/(a2–a1)
    let m2 = (d2–d1)/(c2–c1)
    let k1 = b1–m1*a1
    let k2 = d1–m2*c1
    let x = (k2–k1)/(m1–m2)
    let y = m1*x + k1
    print "\put {$\bullet$}  at ";
    print using "###.##  ": x,y
loop
end
```

For more-challenging problems that I faced in computing figures, see appendix 1.

CONCLUSIONS

Mine was a little problem, with a simple solution, but it illustrates several important points:

1. This was the most standard of problems—find the intersection of two lines, given two points on each line. The context for solving it was different—an algorithm in a computer language instead of a paper-and-pencil algorithm. When the emphasis is on how to effectively compute the answer with the tools available, creativity will always be needed in even the most elementary mathematics. Therefore, a primary goal of mathematics education should be to prepare students to be flexible problem solvers. They will have to create new algorithms or at least modify old ones.

2. What is easy, and what is hard, depends on the tool being used. PICTEX makes it easy to draw intersections but harder to mark them. Suppose I had been using Geometer's Sketchpad instead. Then marking an intersection would have been easy. Something else would have been hard. (In this instance, the figures would have been external to the TEX code and our electronic files would not have been "device independent.") Or compare PICTEX as a tool with straightedge and compass. With those traditional tools, we can use traditional geometric constructions to determine, say, where *E* has to be. With the different tools of PICTEX and BASIC, constructing the figure was an entirely new problem (and lots of fun).

3. People must be able to express solutions in many forms: symbols on paper, symbols in computer languages, button sequences for calculators, and so forth. Notice, for instance, that mathematical language looks different on computers than on paper; there are no subscripts or fraction bars in computer input. In fact, different computer languages have rather different syntaxes, and we must be able to translate back and forth among them as well as to them. Perhaps someday we will be able to write standard paper-and-pencil mathematical syntax on a pressure pad and all computer programs will understand it. But for the foreseeable future, the number of different forms for expressing mathematics will proliferate.

4. Single formulas are not so important anymore. The appropriate solution was a procedure, not a formula, yet out of habit I was hung up on getting a formula, since I knew one existed (see note 3 in appendix 2).

WHAT IS AN ANSWER?

Several of my conclusions involve the same idea: the concept of "answer" is changing. A good answer is easy to understand and easy to use, as well as somehow elegant. But what is easy to use depends on the tools of the time, and what is easy to understand depends on what tools and symbols people are used to. If algorithms can be evaluated quickly and people are used to reading them, then a longer algorithm may be a better answer than a shorter formula. In the case study, the algorithm was just an unpacking of the formula, but an algorithm may not depend on a formula at all and still be a better answer. This is the gist of the following words from a speech by John Kemeny (1983, pp. 204–5):

> I think mathematicians have tremendous pre-occupation with formulas, because they love finding answers "in closed form". Consider the simplest kind of integration problem, say,
>
> $$\int_0^{13} e^x dx.$$
>
> It's the kind of problem you put on a freshman calculus exam so that every student should get at least one problem right. Everybody knows how to do it. The answer is $e^{13} - 1$. Suppose someone came along and said, "Why didn't you use numerical integration?" You'll say that is absolutely mad. Here is a closed form solution, and you can get the exact answer in two minutes; in 10 seconds if you are good. Why in heaven's name would you want to use numerical integration to get an approximate answer? So my question is, if that's the exact answer, please tell me what it is to one significant figure....
>
> Next you'll argue that it is still important that $e^{13} - 1$ is exact. You know, for years I accepted that, until one day I woke up, and it hit me that the original integral
>
> $$\int_0^{13} e^x dx$$
>
> is also exact! I do not mean that as a joke—I mean that as a deep remark about mathematics, one that I overlooked for a large number of years. Now the question is, if you have two forms that are exact, why is one preferable to the other one? And if you think about that particular example, I think it's very easy to reconstruct the reason. We did not have computers but had tables of e^x. Even today there may be some advantage; you may be able to use a pocket calculator instead of a computer on $e^{13} - 1$. But the numerical integration takes less than a second on a computer.... [I]f you put in there $x^2 \sin x\, e^x$ I've got absolutely no doubt whatsoever [what is the better answer]. And it's still less than one second on the computer, and it's the same algorithm that works.... The lesson in this particular case is that mathematicians have got to change their way of thinking from exclusive emphasis on formulas to algorithms.

Wilf (1982) made the same point with respect to counting problems. A good listing algorithm, working directly from the problem statement, may be faster for counting the items in question than evaluating a formula with multiple sums, products, combinations, and so on. If so, the problem statement is a better answer than the formula!

Barwise and Etchemendy (1991) have proposed an even more striking change in what constitutes an answer. Given humankind's very high absorption rate for visual information and given the power of current computers to produce graphics and provide graphical user interfaces, then a picture or a movie may be a better answer than anything written in symbols. Quite a proposal when not long ago many mathematicians eschewed any answer that depended on a picture!

What do you think is a good answer?

REFERENCES

Barwise, Jon, and John Etchemendy. "Visual Information and Valid Reasoning." In *Visualization in Mathematics,* edited by Walter Zimmermann and Stephen Cunningham, pp. 9–24. MAA Notes no. 19. Washington, D.C.: Mathematical Association of America, 1991. Reprinted in *Philosophy and the Computer,* edited by Leslie Burkholder, pp. 160–82. Boulder, Colo.: Westview Press, 1992.

Berzsenyi, George, and Stephen B. Maurer. *The Contest Problem Book V.* Washington, D.C.: Mathematical Association of America, 1997.

Hart, Eric W. "Discrete Mathematics: An Exciting and Necessary Addition to the Secondary School Curriculum." In *Discrete Mathematics across the Curriculum, K–12,* 1991 Yearbook of the National Council of Teachers of Mathematics, edited by Margaret J. Kenney, pp. 67–77. Reston, Va.: National Council of Teachers of Mathematics, 1991.

Kemeny, John. "Finite Mathematics—Then and Now." In *The Future of College Mathematics: Proceedings of a Conference/Workshop on the First Two Years of College Mathematics,* edited by Anthony Ralston and Gail S. Young, pp. 201–8. New York: Springer-Verlag, 1983.

Maurer, Stephen B. "Two Meanings of Algorithmic Mathematics." *Mathematics Teacher* 77 (September 1984): 430–35.

———. "What Are Algorithms? What Is Algorithmics?" In *The Influence of Computers and Informatics on Mathematics and Its Teaching,* edited by Bernard Cornu and Anthony Ralston. Science and Technology Series no. 44. Paris: United Nations Education, Scientific, and Cultural Organization, October 1992.

Maurer, Stephen B., and Anthony Ralston. *Discrete Algorithmic Mathematics.* Reading, Mass.: Addison-Wesley Publishing Co., 1991.

National Council of Teachers of Mathematics. *Curriculum and Evaluation Standards for School Mathematics.* Reston, Va.: National Council of Teachers of Mathematics, 1989.

———. *Discrete Mathematics and the Secondary Mathematics Curriculum.* Reston, Va.: National Council of Teachers of Mathematics, 1990.

Wilf, Herbert S. "What Is an Answer?" *American Mathematical Monthly* 89 (May 1982): 289–92.

APPENDIX 1: EXTENSIONS

Algorithm 1 is not the only program I wrote for the contest book. Here are two more-challenging problems. You might like to give all three problems to your students.

Problem 2. Find the coordinates of the intersection of a line and a circle when the line is given by two points and the circle by the center and radius.

I needed an algorithmic solution to problem 2 to find *T* in figure 3.2. The center of the circle is the known centroid of the equilateral triangle, and thus the radius also was easily computed by hand. The center is a known point on *OT*. Since line *OT* is parallel to side *AC*, a second point on *OT* could easily be found.

Problem 3. Determine the coordinates of the corners of a little right-angle sign, given the point where the perpendicular lines meet, the direction of one of the perpendicular lines, and the desired side length of the sign.

Completing figure 3.2 required the solution of problem 3 as well.

APPENDIX 2: NOTES

1. In figure 3.1, it is possible to mentally compute the coordinates of *F* if the right theorem is known. There is a theorem that states the ratio AF/AD in terms of BD/BC and AE/AC. It turns out that $AF/AD = 1/2$ in this case, so $F = (40, 22.5)$. And in other cases I had to deal with, other special methods could surely be devised. But the point in algorithmic thinking is to avoid repeated special-case solutions, no matter how ingenious.
2. The program displayed in this paper is not quite the program I used in editing the problems book. For instance, in going from m_1 and m_2 to x, I didn't define k but rather a slightly different intermediate quantity less in keeping with traditional notation. But the idea—never actually complete any algebraic substitutions—was the same.
3. The change in emphasis from formulas to procedures involves an interesting historical about-face. Although the word *algebra* derives from the title of a ninth-century Arabic text (*Hisab al-jabr wal-muqabala*), the algebra done at that time is very different from what we do now. The problems were similar to those of today, for instance, solving linear and quadratic equations, and the method was similar—do the same thing to both sides of an equation until the unknown is isolated. But there were no formulas. There weren't even any symbols. Everything was done in words. A solution was a procedure, exemplified by several examples, not a one-line formula. So, in moving from algebra to algorithms, in some sense we are moving back to the original algebra. The about-face is not complete. Our algorithms do contain symbolic expressions. But the solution need not be a single formula.

4

Algorithmic and Recursive Thinking

Current Beliefs and Their Implications for the Future

Tabitha T. Y. Mingus

Richard M. Grassl

Historically, the use of algorithms has been emphasized in the mathematics curriculum at the elementary and secondary school levels. In fact, the dominance of algorithmic procedures has caused many people to define knowledge in mathematics as competence in executing prescribed algorithms. The current reform movements are de-emphasizing the importance of algorithms in favor of problem-solving approaches, the conceptualization of mathematical processes, and applications of mathematics in real-world situations. Each of these areas of emphasis, however, relies heavily on algorithmic or recursive thinking. This situation leaves the educational community with an apparent conundrum: How do we reconcile the decreased emphasis that current educational reforms place on algorithmic processes and isolated computation with the fact that algorithmic thinking provides the formal structure for mathematical growth and understanding? In this paper we address the following questions in an effort to provide a foundation for resolving that apparent conundrum: (1) What do teachers and students believe algorithms are in contrast to how a mathematician, mathematics educator, or computer scientist would define an algorithm? (2) What are algorithmic and recursive thinking and how do they differ from the use of algorithms? (3) In what ways should algorithmic and recursive thinking be incorporated into the secondary school mathematics classroom and how does this benefit the teacher and student?

WHAT ARE ALGORITHMS?

We decided the best way to get a good picture of what students and teachers think algorithms are was to ask them. We surveyed approximately 40 middle and high school in-service teachers and 60 college students. The college students surveyed were preservice elementary school teachers enrolled in a mathematics content course and mathematics majors or minors in an introductory discrete mathematics course, many of whom were also preservice teachers. We also posed problems to 230 middle and high school students to explore their use of algorithms, algorithmic thinking, and recursive thinking. The results were at times surprisingly homogeneous. Each of the groups surveyed provided insightful answers, but rarely on the same questions. The questions from all three surveys appear in the appendix.

The first question on the survey was "What is an algorithm?" The students enrolled in the mathematics content course for elementary school teachers possessed a very limited view of an algorithm; their definitions were terse and simplistic when compared with those of the discrete mathematics students and in-service middle school and high school teachers. This difference may be attributed to the mathematical maturity of the discrete mathematics students and the years of classroom experience of the in-service teachers. The in-service teachers' use of slightly more-simplistic terms than those used by the discrete mathematics students to describe an algorithm may be a function of the in-service teachers' ability to communicate their understanding effectively to students who do not yet possess the language needed to comprehend abstract ideas—a skill put to the test every day spent in the classroom.

Of the thirty-seven responses received from the preservice elementary school teachers, thirty described an algorithm as "a step-by-step procedure used to solve a problem." The remaining seven participants described an algorithm as "a way to do a problem." In contrast, those enrolled in the discrete mathematics course possessed a more detailed conception of an algorithm. Of these twenty-two responses, about half stated that an algorithm was essentially "a step-by-step process with specific input and output designed to solve a problem." The remaining discrete mathematics students described an algorithm much in the way the mathematics education students did. Their definition, "a step-by-step process," omitted the essential components of input and output and the fact that the process is designed to solve a specific problem.

It is clear that the elements contained in the preservice and in-service teachers' responses embrace the essence of how an educator trained in mathematics education, mathematics, or computing science would define an algorithm. The following definition is a more complete description of an algorithm:

An *algorithm* is a computational recipe for the systematic execution of a procedure designed to solve a specific problem that maintains the following characteristics:

1. Input data along with a finite set of instructions are given.
2. A computing agent reacts to the input and instructions and carries out the steps.
3. Intermediate results are stored and used.
4. The computation is carried out in a discrete, stepwise fashion.
5. The computing agent interprets the set of instructions in such a way that computation is carried out deterministically, without resort to random methods.

WHAT ARE ALGORITHMIC THINKING AND RECURSIVE THINKING?

When asked to explain the difference between recursive thinking and algorithmic thinking or whether they saw any connections between problem solving and algorithmic thinking, many of the respondents either used the term *algorithmic thinking,* although their description was actually that of an algorithm, or referred only to algorithms in their responses.

It is important to establish working definitions of algorithmic thinking and recursive thinking, since they will serve as the basis for the discussion in the remainder of this paper.

Algorithmic thinking is a method of thinking and guiding thought processes that uses step-by-step procedures, requires inputs and produces outputs, requires decisions about the quality and appropriateness of information coming in and information going out, and monitors the thought processes as a means of controlling and directing the thinking process. In essence, algorithmic thinking is simultaneously a method of thinking and a means for thinking about one's thinking.

Recursive thinking is a never ending, step-by-step method of thinking in which each step is dependent on the step or steps that come immediately before it. Recursive thinking is iterative and self-referential.

Perhaps the most familiar model of algorithmic thinking is Polya's four-step model. In his model of problem solving, Polya describes a set procedure, or algorithm, for attacking any problem. The first step is to understand the problem, which is the stage at which the problem solvers must determine what inputs are available and the quality and appropriateness of the inputs at their disposal. The second step is to devise a plan, possibly an algorithm within this algorithm. At this point the problem solvers devise a step-by-step procedure to be implemented, determine what input is needed and what output would be produced, and establish guidelines for what an acceptable output would be. The third step is to implement the plan. Here the problem solvers are not only using the procedure they devised to solve the problem, but they are also cross-checking the information the procedure is producing to see if it is appropriate. They are monitoring their own thoughts to ensure they are executing their plan appropriately and that their plan is achieving their goal. If the plan is found to be defective or ineffective, the problem solver must then return to one of the previous steps in order to devise a better plan. The final step is to look back and extend the problem. It is during this final step that the most learning can occur. The problem solvers look at their overall thinking process and determine where it was effective and where it was not and why. They determine which of the methods they developed during the process may prove to be useful in future situations. They may identify other similar problems that can use the same tools or may perhaps connect the methods and content they learned in this problem-solving situation to problems they have solved before by using other methods.

An example of a problem involving algorithmic thinking that we asked several middle and high school students to tackle is the traveling salesperson problem. The students were asked to find the best route for a salesperson to take when visiting four different cities out of a given set of six cities. Solving this problem is a good example of an *algorithmic process* as opposed to a recursive process. Students must first select which four cities they wish to tour, label the six paths, compute the length of each path, and then as a final step in the process, select a path requiring the minimum distance.

Recursive thinking is very algorithmic in the sense that it proceeds in a step-by-step manner. However, the iterative and self-referential nature of recursive thinking distinguishes it from algorithmic thinking. Recursive thinking builds on itself continuously. As one preservice teacher responded, "Algorithmic thinking has an end, a stopping point, while recursive thinking never ends."

We asked the secondary school students involved in the survey the following question to encourage recursive thinking: Suppose you are given 1×1 bricks and 1×2 bricks to construct a $1 \times n$ path. How many different paths can you construct? This question provides a nice example of how algorithmic thinking forms a precise *recursive technique* for determining the number of paths of various lengths. A step-by-step process needs to be devised that builds on having paths of lengths 1 and 2 to start with; a finite set of instructions, consistent with the goal of forming a path of a certain length, needs to be constructed; and on the basis of observed data, a leap (with some reasonable certainty) to the general recursion will follow.

An example of the power of recursive thinking occurs in the use of constructivist techniques in the classroom. Constructivism requires learners to access their previous experiences, build on them through interaction with their teachers and peers, combine their new experiences with their old ones to formulate new knowledge, and then use that new knowledge as a basis on which additional knowledge can be built. Learners must actively engage in the process of building up their knowledge by using previous knowledge as the foundation. Thus the process relies on both the inputs and the outputs available, unlike an algorithm, which relies only on the initial input value. Kieren and Pirie (1991) use recursion as a metaphor in describing how children understand and perhaps construct their own mathematical knowledge. They list the following as the important characteristics of recursive thinking: (1) the process is self-referencing; (2) it involves states that are different from one another; however, the structures of the states are similar; and (3) each level of the process builds on the previous level and is then incorporated as a state that can be called on later to serve as a foundation (p. 79). These three characteristics are the keys to understanding the nature and power of recursive thinking.

IS PROBLEM SOLVING LINKED TO ALGORITHMIC THINKING?

When asked on the survey, "Is problem solving linked to algorithmic thinking?" many of the respondents saw an immediate and direct connection between problem solving and the use of algorithms during the problem-solving process. An example of such a response is, "Problem solving is defining a problem and the possible paths from the problem to the solution. I think that the paths are the algorithms...." However, a second and less apparent connection between problem solving and algorithms is highlighted in the following responses:

> Once an algorithm for solving a problem given specific inputs is formulated, the algorithm can be generalized and applied to a variety of similar problems and used in conjunction with other algorithms to tackle more difficult problems.
>
> Problem solving is the means by which students develop algorithms.

Here we can see that the respondents believe that algorithms are created during the problem-solving episode as a means of devising and implementing a plan to answer the question. Both of these connections were related to specific algorithms (either known or created) being used in the problem-solving process, not to the connection algorithmic thinking had to problem solving. The last connection that was highlighted by the respondents was more general and referred to the algorithmic-thinking process instead of to the use of algorithms. The respondents viewed problem solving as a generalization of algorithmic thinking and algorithmic thinking as a critical factor in successfully developing, executing, and directing the problem-solving process, as illustrated by the following comments: "Algorithmic thinking can be a part of problem solving." "Sensing a plan for problem solving often requires algorithmic thinking." The most startling aspect of the responses to the survey was the consistent use of a mechanical, recipe-like description of algorithms in place of algorithmic thinking, which is broader and much more powerful than a single, simple algorithm.

Is There a Difference between Algorithmic Thinking and Recursive Thinking?

A similar misconception was apparent in the responses to the question "Is there a difference between algorithmic thinking and recursive thinking?" As in the previous question, there was confusion between the terms *algorithmic thinking* and *algorithm* and between the terms *recursive thinking* and *recursion.* Both *algorithm* and *recursion* refer to specific methods or relations that are used in solving a specific mathematical problem. Equating a single algorithm to algorithmic thinking is akin to saying lecturing is the same as teaching. Lecture is but one means by which we try to teach students, just as an algorithm is but one tool with which we think. The following are examples of this common misconception from our respondents:

> Recursive thinking depends on input being the output from before, whereas algorithmic thinking does not depend on an output. Recursive thinking builds off the previous answer while algorithmic may build off the previous answer but it doesn't always need to happen.

> The difference between algorithmic thinking and recursive thinking is that an algorithm can jump forward or backward within the procedure and recursion must use previous information to continue.

The consistent juxtaposition of the tool (recursion or algorithm) with the thinking process did not diminish the intuitive understanding of the subtle differences between algorithmic thinking and recursive thinking. The dependence of recursive thinking on prior knowledge and the iterative process of building knowledge on knowledge is clear from the comments. The traveling salesperson problem also illustrates one of the important differences between algorithms and recursions. In the process of determining the minimum path, only the input was needed, not the output. In contrast, the output at each stage of the recursive process used to determine the number of brick paths is used as input in the next stage.

Using Algorithms and Algorithmic Thinking to Enhance and Assess Students' Understanding

One way that algorithms and algorithmic thinking can be incorporated into the curriculum is by teaching students how to use algorithms as tools to construct their own mathematics rather than as ends in themselves or how to analyze the algorithms used by others to solve problems. In the 1991 National Council of Teachers of Mathematics (NCTM) Yearbook on discrete mathematics, Gardiner put forth that "the educational value of much simple discrete mathematics lies precisely in the fact that it forces students to *think* about very elementary things" (p. 12). One could say the same about algorithms. Frequently, simple algorithms describe processes that are (or have become) automatic to both the teacher and student. All too often small bugs can creep into students' execution of the algorithm and cause them to make consistent and perplexing errors. (For the middle and high school students we surveyed, this was true of the division algorithm. The students mistakenly adopted the practice of placing the remainder after a decimal point instead of as the numerator of a fraction.) Students may perform the algorithm seemingly perfectly; however, given certain inputs they run into errors. By requiring students to examine the processes carefully that they are going through and to verbalize them on paper, the teacher (or other students) can follow the students' algorithms and find the hidden bugs in their thought processes.

Gardiner (1991) goes on to state in regard to teaching algorithms that "algorithms tend to become obsolete sooner rather than later. The algorithms themselves are therefore of limited interest. What matters most for the students' long-term mathematical education are the mathematical ideas behind the design, construction, and analysis of such algorithms and the experience of applying some of these ideas to devise and improve their own simple algorithms" (p. 13). Students are often told they must take mathematics in high school because it will teach them how to think mathematically. What better way to learn how to think mathematically than to dissect and examine how a mathematician thinks? Algorithms are one means by which we can look into one anothers' minds (and the minds of mathematicians before us) and see what thought processes we are using and why. After considering some simple traditional algorithms, the next step for students is to create their own algorithms and compare and contrast them with those of their peers. The process of analyzing and creating algorithms requires the student to use algorithmic thinking, which will never become obsolete. This process gets students actively involved in creating mathematics—mathematics of which they have joint ownership.

In the same yearbook (NCTM 1991, pp. 195–96), Maurer and Ralston state:

> Not only does the algorithmic approach allow the use of realistic problems, it also forces you to generalize, that is, to consider all aspects of a problem, not just special cases.... Being forced to generalize may not sound like an advantage, but mathematical knowledge is only really useful when it allows a class of problems rather than specific cases to be understood and attacked.... Writing general algorithms requires more attention to mathematical notation than has been typical in secondary school.

Although some may argue with Maurer and Ralston's statement about when mathematical knowledge is "really useful," the important point to glean from this passage is that mathematics is a powerful tool and competency in the use of that tool provides an individual with a great deal of power to bring to bear on any problem he or she may face in life. The primary goal in the mathematics classroom is to develop students' mathematical power. Although the NCTM (1989) *Curriculum and Evaluation Standards for School Mathematics* is nearly a decade old, its definition of mathematical power still rings true today. Mathematical power is defined as "an individual's abilities to explore, conjecture, and reason logically, as well as the ability to use a variety of mathematical methods effectively to solve nonroutine problems" (p. 5). Properly integrated into the curriculum, algorithms can play an important part in developing computational skills, reasoning ability, and a playful attitude in our students. Without these, students will have little success at solving or even wanting to solve nonroutine problems.

When we asked our preservice and in-service teachers how they would or do "use algorithms in your classroom," we found that many of them saw algorithms as a means for teaching computational skills or as a way to allay the fears of students who are math anxious by giving them a set procedure to memorize and implement. These uses are appropriate for some students. Others saw the potential of using algorithms as a basis for teaching higher-level reasoning. The following are some of their suggestions:

> I would give a simple example of something the students already know. Long division for example, then explain what an algorithm is and what is required to make an algorithm. Then I would help them create their own.
>
> I would teach the students how to perform some of the different algorithms and allow time for discussion.
>
> In geometry, my students read algorithms (and follow them) to do constructions. They then write their own algorithms and trade with a partner. I also have them follow their own algorithms a week after writing them and make necessary revisions.
>
> By taking a problem from an abstract form to a concrete form and building step-by-step from the concrete, I can teach a student WHY a problem works, which is important for long-term learning.

Just as simple algorithms can reinforce arithmetic skills, slightly more-complex nonroutine problems (which are contrived so they behave nicely) can be used to develop needed problem-solving strategies and a sense of self-confidence so students are then willing to be adventurous and to tackle more unwieldy real-world problems. Also, there is incredible value in doing mathematics for the sake of doing mathematics, and as teachers we need to emphasize that to our students. The utility of mathematics is a wonderful thing, but if we always judge the value of a mathematical activity on the basis of its current utility, we may block today's discovery of tomorrow's tools.

Assessment is a critical component in both teaching and learning mathematics. The in-service teachers were asked, "How do you assess algorithmic thinking?" Their comments revealed numerous connections to teaching proofs, problem solving, writing, cooperative learning, and verbalizing concepts:

> I would assess student understanding through problem solving.... I might ask students what is the maximum number of intersections of 2, 3, 4, 5, 6 lines in the plane and then have them explain how they know they have the maximum.... The process they describe is their algorithm for finding maximum intersections.
>
> I would give problems where they find the greatest common multiple or solve an equation and have them show their steps.
>
> Through activities and presentations of the activities and projects which are designed to encourage algorithm/problem solving. Looking for patterns and rules to explain the data they collect.
>
> Students write algorithms, trade and follow someone else's. Also follow their own. Also, in class discussions, volunteers "talk" us through their algorithm until we do (or hopefully, do not) hit a problem. As students talk, I write and follow their steps so I can demonstrate any problems—kids revise as we go.

What Role Will Algorithms Play in the Classrooms of the Future?

The in-service teachers surveyed were quite emphatic about their perceptions of the "connections between teaching algorithms and the NCTM Standards":

> The Standards emphasize the process, how we think, how we figure out—looking for patterns—patterns lead to rules and this to algorithms. We need to teach students how to develop their own algorithms and break away from traditional algorithm instruction. Students should identify their own steps and ways for solving problems. Teaching algorithms helps kids to organize their thoughts and to see the importance of this organization.

These same teachers were equally strong in their thoughts on how the use of algorithms affects students' attitudes about mathematics:

> The use of algorithms allows students to look at math as a process rather than as a question-answer type activity. They feel better about themselves and the class because they can do a problem no matter how complex; they can choose from their "toolbox." Algorithms provide a comfort zone for some students and encourage students to pursue better ways as they get comfortable with them. My students enjoy finding their own methods and fall into important mathematical justifications when they write their own algorithms.

A similar question was asked of the college students: "If you are familiar with the NCTM Standards, do you see connections between teaching algorithms and the Standards?" Since the majority of these students were preservice teachers, we expected them to be conversant with the expectations contained in the *Standards*. This was not so. Of the twenty-two students who were mathematics majors, thirteen did not respond to the question at all and the remaining nine stated they were "not familiar with the NCTM or the Standards." The thirty-seven preservice elementary school respondents were equally unfamiliar with the NCTM *Standards*. Of them, thirteen did not respond, twenty were "not familiar with the NCTM or the Standards," one saw no connection to the Standards at all, and three saw links between algorithms and the Standards. The following is an example of the beliefs of preservice teachers who were familiar with the Standards:

> Yes, the Standards want kids to see patterns. Using algorithms develops a pattern for solving problems. With knowledge of algorithms learning can be placed in the hands of the learners.

Clearly, there is still much work to be done in discussing educational issues like the NCTM *Standards*, constructivism, cooperative learning, and the like, so that preservice teachers are well prepared when they enter the classroom for the first time.

In a classroom lesson on using and creating algorithms and recursions, we gave middle school and high school students four problems to solve over a two-day period. We found that as a result of teaching students about algorithms, recursions, algorithmic thinking, and recursive thinking, they began to think in a more logical, step-by-step manner. The students were encouraged to write out their thinking, which allowed us to identify and correct misconceptions that prevented their progress toward solutions of the problems. As they began articulating their thoughts, it became necessary for them to develop better notation and language to communicate their ideas to the students with whom they were working on the problems. The process of establishing language and notation for their thoughts deepened their understanding and ownership.

Incorporating instruction on algorithms and algorithmic thinking into the secondary school curriculum will encourage students to grow as problem solvers. However, we need to provide students with more opportunities to tackle tougher and less routine problems. We must also provide them with the tools to enable them to create their own algorithms and thus to construct their own mathematics. There is nothing quite as inspiring as ownership.

REFERENCES

Gardiner, Anthony D. "A Cautionary Note." In *Discrete Mathematics across the Curriculum, K–12*, 1991 Yearbook of the National Council of Teachers of Mathematics, edited by Margaret J. Kenney, pp. 10–17. Reston, Va.: National Council of Teachers of Mathematics, 1991.

Kieren, Thomas, and Susan Pirie. "Recursion and the Mathematical Experience." In *Epistemological Foundations of Mathematical Experience,* edited by Leslie Steffe, pp. 78–101. New York: Springer-Verlag, 1991.

Maurer, Stephen B., and Anthony Ralston. "Algorithms: You Cannot Do Discrete Mathematics without Them." In *Discrete Mathematics across the Curriculum, K–12,* 1991 Yearbook of the National Council of Teachers of Mathematics, edited by Margaret J. Kenney, pp. 195–206. Reston, Va.: National Council of Teachers of Mathematics, 1991.

National Council of Teachers of Mathematics. *Curriculum and Evaluation Standards for School Mathematics.* Reston, Va.: National Council of Teachers of Mathematics, 1989.

APPENDIX

Survey Questions Given to In-Service Teachers

1. What is an algorithm?
2. Is there a difference between algorithmic thinking and recursive thinking? If yes, what is it?
3. How do you use algorithms in your classroom? Illustrate with examples.
4. How does the use of algorithms affect students' attitudes about mathematics?
5. Is problem solving linked to algorithmic thinking? Explain, and give examples if possible.
6. How do you assess algorithmic thinking? Give specific examples if possible.
7. Do you see connections between teaching algorithms and the NCTM Standards? If so, please describe them.
8. What importance, if any, do you attach to the teaching of algorithms? Please elaborate.

Survey Questions Given to Preservice Teachers and College Mathematics Students

1. What is an algorithm?
2. Is there a difference between algorithmic thinking and recursive thinking? If yes, what is it?
3. How do you as a student or a future teacher use algorithms in your classroom? Illustrate with examples you have encountered or used.
4. If you are a preservice teacher, would you teach algorithms in your classroom and how would you do so?
5. Is problem solving linked to algorithmic thinking? Explain, and give examples if possible.
6. If you are familiar with the NCTM Standards, do you see connections between teaching algorithms and the Standards? If so, please describe them.

Questions Given to Secondary School Students

1. Describe the process you use for dividing one number into another number.
2. Describe how you would change a number in base ten into base two.
3. Suppose a salesperson is going on a business trip and must travel to four of the following six cities: Cincinnati; Denver; Atlanta; Washington, D.C.; Seattle; and San Francisco. Select four cities and determine the best (shortest) route for the salesperson to travel. (Students were given a map with distances between cities listed.)
4. Suppose you are given 1×1 bricks and 1×2 bricks to construct a $1 \times n$ path. How many different paths can you construct? (Note: A 1×3 path that consists of a 1×2 brick followed by a 1×1 brick is different from a path that consists of a 1×1 brick followed by a 1×2 brick.)

5

Teaching Mental Algorithms Constructively

Alistair McIntosh

IN EVERYDAY life we use mental computation far more than formal written arithmetic. Even in 1957, before electronic calculators were universally available, adults used formal written computation for only 25 percent of their calculations (Wandt and Brown 1957); the rest they computed mentally. And this is true not only of adults. For example, Carraher and Schliemann (1985) have shown that children both tend to prefer informal mental strategies to formal written algorithms and are more proficient in their use.

I regularly ask groups of elementary school teachers how much of the time in their mathematics classes they devote to teaching formal written algorithms. They say that depending on the class they teach, they spend between 50 percent and 95 percent of their mathematics time teaching formal written algorithms; and yet they agree that these are seldom used, particularly in an age of electronic calculators and computers, and that outside the classroom people are much more likely to need and use mental-computation skills.

For this reason alone we should shift our emphasis to the teaching of mental computation. How should we do this? Traditional pedagogical methods would suggest that we should look for the best mental-computation algorithms and teach them. I suggest, however, that doing so may be counterproductive.

The differences between standard written algorithms and mental algorithms have been brilliantly analyzed by Plunkett (1979). He points out that although formal written algorithms have the advantage of providing a standard routine that will work for any numbers—large or small, whole or decimal—they also have disadvantages in that they do not correspond to the ways in which people tend to think about numbers and they discourage students from thinking about the numbers involved or from exercising any active choice or initiative while carrying out the computation. Mental-computation strategies are quite different: they are flexible and can be adapted to suit the numbers concerned; they involve a definite, if not conscious, choice of strategy based on considering the numbers involved; and, almost always, they require understanding. I have found one exception to the last rule, and it is highly significant, as I shall discuss later.

But the most important distinction between normal written and mental-computation algorithms at present is exemplified by the following: Suppose we say to a class, "What is 36 + 79? Do it with pencil and paper." Then, by and large, we will expect every child in the classroom who gets the problem right to do the same thing: "6 and 9 are 15; put down the 5 and carry the 1; 3 and 7 are 10, and 1 more is 11. The answer is 115." But ask the same class to do the same calculation mentally and the situation alters radically. In my experience and that of colleagues who have been asking children similar questions over the past ten years in a variety of classrooms, we would expect to hear any or all of the following:

"3 and 7 are 10; 6 and 9 are 15; that's 115."

"30 and 70 are 100; 6 and 9 are 15; that's 115."

"36 and 80 are 116; less 1 is 115."

"36 and 70 are 106, and 9 is 115."

"79 and 6 are 85, and 30 is 115."

"79 and 21 are 100; 36 less 21 is 15; 100 and 15 are 115."

A number of students, particularly those from classrooms in which formal written algorithms have been heavily emphasized, will screw up their eyes, raise their hand as though writing in the air in front of them, and say, "6 and 9 are 15; put down the 5 and carry the 1; 3 and 7 are 10, and 1 more is 11. The answer is 115."

All these methods work, all of them are more or less efficient, they rarely appear to have been taught, and almost all take account of the particular numbers involved, as opposed to being automatically applied. We can tell this because when we ask children to do another similar calculation, for example, 25 + 83, they often use a quite different mental strategy.

Almost all the children we have interviewed have, one way or another, acquired a range of strategies for performing mental computations, but the less confident or competent are particularly handicapped by the lack of specific classroom support in acquiring more-efficient and more-reliable strategies. Moreover, in asking seventy-two children in grades 2 to 7 in three different schools to explain their strategies for a total of more than three thousand mental computations, we found that the same range of mental-calculation strategies was eventually available to almost all the students: but the more able children acquired them up to two years earlier.

The students started by using the most primitive strategies based on counting forward and back by ones. They then progressed to a range of more-sophisticated strategies—counting forward and back in larger numbers, using number properties such as commutativity, bridging 10 or multiples of 10 (with 38 + 7, for example, 38 and 2 is 40 and 5 more is 45), building on known facts, compensating (e.g., 28 + 27 is the same as 30 + 25), using place value to split up numbers, and so on. We are not talking here about highly intelligent children, whom we deliberately excluded from our research, but the general run of ordinary children.

These children have powers of mental computation that they develop themselves without support. The heart of all these self-devised algorithms is that the child tries to turn a difficult calculation into an easy one: 28 + 27 is hard, 30 + 25 is easy; 9 × 17 is hard, 10 × 17 – 17 is easier. The more able children develop their ability to devise algorithms and so make life easy. Unfortunately, the less able children are less capable of helping themselves and so are left with much more difficult tasks. Try starting with 28 and adding on 27 in ones without error, and you will appreciate their problem.

We must find ways of helping children in classrooms improve their mental-computation abilities, but it appears that teaching them particular methods may be unhelpful. We are led to that belief because when we interviewed the students, we found only one strategy that had been deliberately taught, either by parents or teachers, to a significant number of children. It was the only strategy the children tried to use without understanding. The strategy is commonly referred to by the stduents as "taking off the zeros." "What is 70 + 20?" "90." "How did you do it?" "I took off the noughts, 7 + 2 = 9; put back the noughts, 90." "Why does it work?" "Don't know." "My mother [or father or teacher] showed me." "How many noughts did you take off?" "Two." "How many did you put back?" "One." "Why?" "Don't know. It just works."

This lack of understanding leads to few errors in addition or subtraction, but it causes problems in multiplication and division. When asked to calculate mentally 150 ÷ 30, most children who have been taught this rule take off the noughts, then "put back" one or both of them, which produces an answer of 50 or 500. Because they have learned to do this algorithm without understanding, they rarely query the answer, even when it is put in a real-life context, for example, ordering thirty-seat buses to take 150 children on an outing.

I recall one child who had no problems with the mental calculation because she approached it in an entirely different way. She was a capable eight-year-old child who had not been taught about removing zeros, and she answered, almost straight away, "5." "Why?" "Well, 30 and 30 is 60, and 30 is 90, and 30 is 120, and 30 is 150. Five 30s." This child understood the problem, thought about the numbers, hadn't been taught how to do the calculation, and used a method she was in control of. In trying to help the other children by teaching them what to do, teachers and parents had quite unconsciously prevented them from using their own number sense.

So what do we do?

We must stop restricting our mental-arithmetic sessions to bursts of short, unrelated calculations in which we emphasize accuracy and speed. If children are given time, they try—often with success—to invent an algorithm. If we emphasize speed, we remove this possibility. So first, don't pressure children by emphasizing speed.

Second, when a calculation has been done, do not concentrate only on whether the students have the right or the wrong answer. Ask at least three or four children to explain how they did the mental calculation, and show interest and enthusiasm for the variety of algorithms presented. Students need to realize that there are many ways of doing a calculation mentally and that there is rarely one "best" method. Of the mental algorithms that children hear discussed, they will use the one that they trust and understand best.

Often you will notice a heightening of interest in a method described by a child, which indicates that the algorithm has been understood and appreciated by several children. At such times "teaching an algorithm" is valid. You might say, "Let's see if we can all use Amanda's method to do these calculations." Making such a suggestion is quite different from imposing a single algorithm from outside. The situation is more like what happens when someone shows off a new toy. A child might ask, "Can I have a go with that?" We all want to try it out a few times to see if we would like it for ourselves. So third, let the students' interest be the springboard for practicing mental algorithms.

Finally, make experiences with mental arithmetic pleasurable for all the children by shifting the emphasis from testing their performance to supporting and encouraging their attempts to think for themselves. Here are three suggestions to get you going. They are based on the work of hundreds of Australian primary school teachers whom we have observed adapting and using these ideas in their classrooms (McIntosh, De Nardi, and Swan 1994):

1. Instead of asking students the answer to a calculation, do the reverse! Say, "I did a calculation, and the answer was 12. What might the calculation have been?" Stick to a simple number—between 10 and 30—so that all the children can engage at their own level. Record all the answers and have the children discuss which are alike. Later you can challenge the students to write down as many ways as they can think of to make 12.
2. Give a calculation, give children plenty of time to work out their answer, establish the answer, then ask, "Who can tell me how he or she did it?" "Do you understand Dan's way?" "Who did it the same way as Dan?" "Who did it a different way?"
3. Students very often deal with abstract calculations in class that they never relate to real life. To help the students make this vital link, give them a calculation, for example, $20 - 7$ or 15×4, let them give the answer, then invite them to suggest situations like the following that incorporate that

calculation: "I had $20, but I spent $7. I have only $13 left." "I have 15 toy cars. All together they have 60 wheels." Often children restrict themselves to stories about money or lollipops, so widen their horizons by asking them to compose a fantasy story or a situation in the kitchen or the zoo.

Activities such as these help your students view the development of their mental-calculation algorithms as an enjoyable, relevant, and approachable challenge. What I am proposing is not new. In fact it is simply elaborating on the advice Warren Colburn gave more than one hundred fifty years ago (Colburn 1970):

> The learner should never be told directly how to perform any operation in arithmetic.... Nothing gives scholars so much confidence in their own powers and stimulates them so much to use their own efforts as to allow them to pursue their own methods and to encourage them in them.

REFERENCES

Carraher, Terezinha Nunes, and Analúcia Dias Schliemann. "Computation Routines Prescribed by Schools: Help or Hindrance?" *Journal for Research in Mathematics Education* 16 (January 1985): 37–44.

Colburn, Warren. "Teaching of Arithmetic." In *Readings in the History of Mathematics Education*, edited by James K. Bidwell and Robert G. Clason, pp. 24–37. Washington, D.C.: National Council of Teachers of Mathematics, 1970. Reprinted from *Elementary School Teacher* 12 (June 1912): 463–80. (Text of an address delivered by Warren Colburn before the American Institute of Instruction in Boston, August 1830)

McIntosh, Alistair J., Ellita De Nardi, and Paul Swan. *Think Mathematically*. Melbourne, Victoria: Longman, 1994.

Plunkett, Stuart. "Decomposition and All That Rot." *Mathematics in School* 8 (May 1979): 2–5.

Wandt, Edwin, and Gerald W. Brown. "Non-Occupational Uses of Mathematics: Mental and Written—Approximate and Exact." *Arithmetic Teacher* 4 (October 1957): 151–54.

6

What Criteria for Student-Invented Algorithms?

Patricia F. Campbell

Thomas E. Rowan

Anna R. Suarez

ALGORITHMS are a part of mathematics; they are necessary and useful. However, the traditional algorithms that all students must memorize should not define elementary school mathematics. We suggest that there should be criteria by which student-invented algorithms can be evaluated and found as acceptable as the traditional ones and that this process will enhance students' learning.

Project IMPACT (Increasing the Mathematical Power of All Children and Teachers) was a National Science Foundation–funded project located in urban schools with diverse populations where low achievement had been typical. Its purpose was to support and enrich students' conceptual development in classrooms by focusing on problem solving. This focus meant that students had to learn to think mathematically and to construct relationships giving order to, and defining patterns across, real-world experiences and problem situations (Campbell and Johnson 1995). In IMPACT, students often solved problems by inventing algorithms on the basis of their interpretations of the problems, their understanding of arithmetic operations, and their representation of numerical relationships.

In IMPACT classrooms, procedures developed by students were usually based on either (*a*) strings of mental arithmetic and ideas incorporating

The research reported in this material was supported in part by the National Science Foundation under Grant Numbers MDR 8954652 and ESI 9454187. The opinions, conclusions, or recommendations expressed in these materials are those of the authors and do not necessarily reflect the views of the National Science Foundation.

number sense or (*b*) manipulative routines using materials or drawings. Teachers assisted students in representing their procedures symbolically first by recording with symbols what the students said and later by guiding the students to record what they said in more efficient ways.

FOSTERING STUDENT-GENERATED ALGORITHMS IN THE CLASSROOM

How does a teacher support a climate where each student's thinking is valued and the teacher's role as a demonstrator of procedures is de-emphasized? A productive technique in Project IMPACT classrooms was to present challenging, accessible problems, allow students to decide how to solve them, and expect students to consider various strategies and to be prepared to share their thinking about the strategies. Sometimes the numbers used in problems were changed to meet differing needs of students. At other times, students were allowed to use numbers of their choice in problems. This assured accessibility and gave teachers an indication of a student's level of confidence with numbers.

There was much communication among students and between the teacher and students. Children who finished quickly often shared ideas while others continued to work. These conversations focused on validating answers and understanding one another's strategies. Classroom instruction that supports student-generated algorithms does not require each child to invent a unique procedure. Indeed, students often decided to use a method invented by another student, saying, "I used Delonte's way." Sometimes invented procedures were really traditional algorithms, or the traditional algorithm was shared after being learned outside school. Teachers sometimes shared the traditional algorithm as "the way I was taught to do it" but emphasized that it was not the only way. The main point was that students were not to use an algorithm unless they could explain it.

Teachers helped students generate and record algorithms by asking questions. Some questions highlighted the decisions students were making with solution strategies: "Do you always start with hundreds? Why?" Other questions highlighted conceptual connections: "Why did you count up? If you wrote a number sentence for this, what operation would it show? Why?" Teachers also asked questions to cause students to consider similarities and differences among procedures. Later, the students began to ask these same questions of one another, and their ability to listen to one another increased tremendously. Students got correct answers, but the answers alone did not provide insight into their understanding. Their understanding was revealed as they discussed their strategies and interacted with their classmates.

MATHEMATICAL CRITERIA FOR ASSESSING STUDENTS' ALGORITHMS

Let's consider a student-generated algorithm for division. A fourth grader was given this problem:

> Four children had three bags of M&M candies. They opened all three bags and shared the candies fairly. There were 52 candies in each bag. How many M&M candies did each child get?

Yolanda started by multiplying 3 × 52 traditionally and got 156. She attempted the traditional division algorithm, but she couldn't remember the procedure. She then drew four circles, wrote "30" beside each, and started making tally marks (see fig. 6.1). She put five tallies in each circle and checked the total by counting (120 ... 125, 130, 135, 140). This routine was repeated as she added two more tally marks to each circle and counted again, moving up to a sum of 148. Finally, she added two more tallies to each circle and arrived at 156. Then she counted the amount tallied and the value she had assigned to one circle and gave her answer of 39.

Fig. 6.1. Yolanda's procedure

This fourth grader solved a two-step problem using an invented procedure based on her understanding of division. Her persistence indicated that this procedure made sense to her. That is very important. However, is her procedure appropriate for continued use? No. Indeed, a year later Yolanda solved and explained the same problem with a more efficient algorithm—the traditional one.

In Project IMPACT, many teachers came to accept student-generated algorithms and to celebrate the kind of understanding that was demonstrated by students like Yolanda. This did not mean, however, that any procedure was acceptable just because a student invented it. Teachers had to consider whether the procedure was (1) *efficient* enough to be used regularly without considerable loss of time and without frustration due to the number of recorded steps required. Student-invented algorithms also had to be (2) *mathematically valid* and (3) *generalizable.* These three criteria became the "standards" for algorithms.

Efficient Procedures

Efficiency is the easiest of the three criteria to apply. If students were drawing pictures, counting tallies, writing lengthy lists of numbers, or just taking an especially long time to work through a procedure, then it was not efficient. When this happened, the teacher provided opportunities for students to move toward something more efficient or helped students develop a more efficient way of recording their mathematical thinking. The teacher might ask the child to explain the procedure again, saying, "This time I will write it down as you explain, so we can understand and remember the steps." This written record would be a more efficient recording. At other times teachers might say, "Would you explain that again, only this time use numbers instead of those tally marks (or drawings)?"

Mathematically Valid Procedures

The issue of validity was not always easy to apply. Consider the two examples of student-invented algorithms shown in figure 6.2. These procedures seem relatively efficient. Are they mathematically valid?

It is not immediately clear whether Dennis's subtraction algorithm is mathematically valid, although in this example it yields the correct answer.

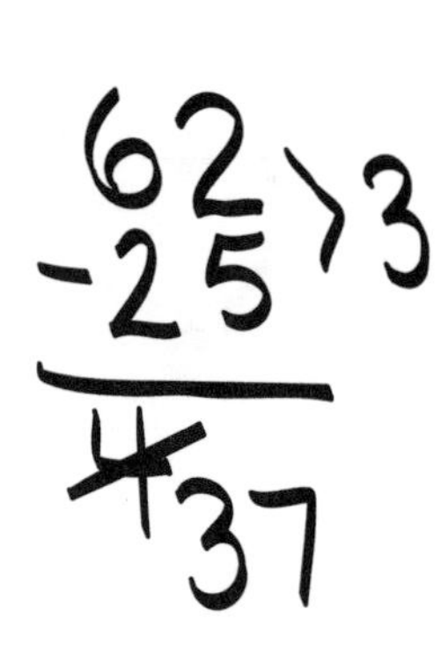

Dennis's subtraction

62.... You have 62. Take 20 away.... Leaves 40. Now you'll .... Then the 2 and the 5. 40 and take 5 away from the ... take 5 away from the 2. Take 2 away from the 5, and then you got 3 away from the 40. Take 3 away from the 40, and that's 37. The answer is 37.

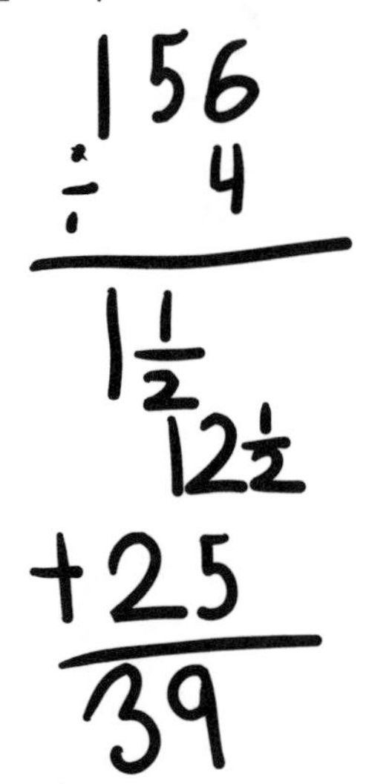

Tyrone's division

6 divided by 4 is 1 and a half. 50 divided by 4 is 12 and a half. 100 divided by 4 is 25. Now add. The answer is 39.

Fig. 6.2. Two student algorithms

The idea of always subtracting the smaller number from the larger is a common misconception for some children who learn the traditional algorithm by rote procedures, particularly if subtraction of multidigit numbers without regrouping is taught before subtraction with regrouping. However, Dennis is not just subtracting smaller from larger in the ones place—he is systematically representing his manipulative actions with base-ten blocks. He first removed the appropriate tens blocks, which research indicates most children do when given the freedom to determine their own strategies (Baroody and Standifer 1993; Madell 1985). Then Dennis needed to remove five ones blocks. He removed the two ones he had, but he still needed to remove three more, the difference between five and two. He subtracted these three from one of the remaining tens. Dennis's efficient procedure is also mathematically valid.

Now consider Tyrone's work. His method of dividing by starting from the ones is intriguing. It also looks efficient, but Tyrone paused for long periods of time as he found each partial quotient. It was not easy for him to determine the fractional part of each partial quotient. Given more practice, though, it might be considered efficient. Is it valid? It is, since it is essentially the partitioning of the dividend, the total, into partial quotients as determined by their place value.

Generalizable Procedures

The final question is to ask if a procedure is generalizable. Can the algorithm be applied to the full range of problems of the type being solved? At first glance, it may seem that Dennis's algorithm is not generalizable, since it may become cumbersome and difficult to apply when used with numbers greater than 99. In figure 6.3, we have applied Dennis's algorithm to a three-digit problem.

Although the procedure is somewhat longer with larger numbers (as is the traditional algorithm), it is generalizable.

Is Tyrone's algorithm generalizable? Consider it when applied to a problem involving larger numbers (see fig. 6.4). Is it efficient enough? Probably

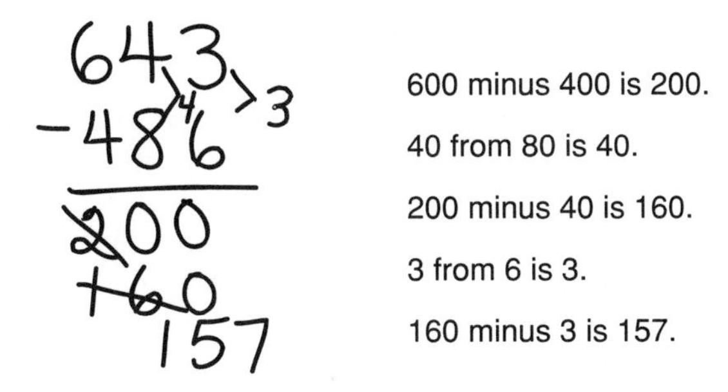

Fig. 6.3. Dennis's algorithm applied to a three-digit problem

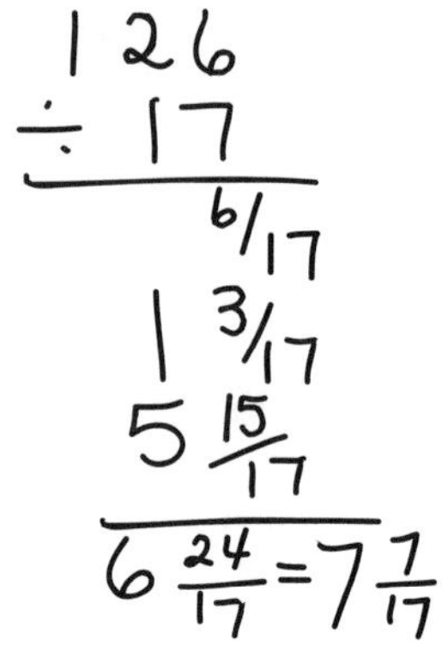

$$\begin{array}{r} 126 \\ \div\ 17 \\ \hline \frac{6}{17} \\ 1\frac{3}{17} \\ 5\frac{15}{17} \\ \hline 6\frac{24}{17} = 7\frac{7}{17} \end{array}$$

6 divided by 17 is 6/17.

20 divided by 17 is 1 and 3/17.

100 divided by 17 is 5 and 15/17.

Add those.

[You get] 6 and 24/17, which is 7 and 7/17.

Fig. 6.4. Tyrone's algorithm applied to larger numbers

not. Tyrone's algorithm has limited generalizability because of the difficulty of finding the fractional parts of the partial quotients. Tyrone did not continue using this algorithm. Had he tried, he would have found a further difficulty when trying to extend the procedure to decimal division. The algorithm has limited generalizability because of its loss of efficiency and its increased difficulty when applied to decimals.

STUDENTS AS MATHEMATICIANS

Sometimes the criteria of efficiency, validity, and generalizability can be applied easily. On other occasions, applying these criteria requires considerable thought. Teachers should expect students to make their thinking clear as they invent procedures. If an algorithm seems to work with a variety of problems, it may be mathematically valid. Determining the validity requires an understanding of why the procedure "works." Efficiency also deserves thoughtful consideration; sometimes unfamiliarity makes an algorithm seem inefficient. At the same time, do not decide that a procedure is valid, efficient, and generalizable just because it works for one or two problems.

Students should be involved in deciding on the validity, efficiency, and generalizability of their own algorithms and those of their classmates. Have students try a suggested algorithm on some problems. These investigations of student-invented algorithms are not intended to lead every child to learn every suggested algorithm. When students are engaged in trying to figure out the meaning of a suggested algorithm, they are becoming more aware of what it means to think mathematically. The kind of thinking that goes into inventing and evaluating procedures for solving problems can be intriguing, rewarding, and stimulating. It is part of the work of mathematicians.

REFERENCES

Baroody, Arthur J., and Dorothy J. Standifer. "Addition and Subtraction in the Primary Grades." In *Research Ideas for the Classroom: Early Childhood Mathematics,* edited by Robert J. Jensen, pp. 72–102. Reston, Va.: National Council of Teachers of Mathematics, 1993.

Campbell, Patricia F., and Martin L. Johnson. "How Primary Students Think and Learn." In *Seventy-five Years of Progress: Prospects for School Mathematics,* edited by Iris M. Carl, pp. 21–42. Reston, Va.: National Council of Teachers of Mathematics, 1995.

Madell, Rob. "Children's Natural Processes." *Arithmetic Teacher* 32 (March 1985): 20–22.

ADDITIONAL READING

Brownell, William A. "When Is Arithmetic Meaningful?" *Journal of Educational Research* 38 (March 1945): 481–98.

Davis, Robert B. *Learning Mathematics: The Cognitive Science Approach to Mathematics Education.* Norwood, N.J.: Ablex, 1986.

Erlwanger, Stanley H. "Benny's Conception of Rules and Answers in IPI Mathematics." *Journal of Children's Mathematical Behavior* 1 (Autumn 1973): 7–26.

Kamii, Constance, Barbara A. Lewis, and Sally Jones Livingston. "Primary Arithmetic: Children Inventing Their Own Procedures." *Arithmetic Teacher* 41 (December 1993): 200–203.

Sowder, Larry. "Children's Solutions of Story Problems." *Journal of Mathematical Behavior* 7 (December 1988): 227–38.

7

The Importance of Algorithms in Performance-Based Assessments

Dominic Peressini

Eric Knuth

UNTIL recently, traditional school mathematics was primarily treated as the development, drill, and practice of algorithmic skills (Cangelosi 1996; Fowler 1994). In an attempt to shift the focus and energies of mathematics teachers toward the more creative and meaningful aspects of mathematics, contemporary reform recommendations de-emphasize paper-and-pencil algorithms (National Council of Teachers of Mathematics [NCTM] 1995, 1996; Mathematical Sciences Education Board 1991; U.S. Department of Education 1995). As an alternative, these documents recommend an increased emphasis on mathematical problem solving, communication, reasoning, and connections. Some teachers (as well as parents, their children, and other community members), however, may conclude that these suggestions indicate that the development of students' algorithmic skills is no longer significant.

This conclusion, however plausible, is not the intended message of these documents. Indeed, algorithms encompass every branch of mathematics and are a necessary part of everyday life. Algorithms not only can represent but also can reveal the complexity of much real-world phenomena. Moreover, the development of algorithmic thinking provides the foundation for students' mathematical power. Steen (1990, p. 7) delineates this utility of algorithms in contemporary mathematics:

> Algorithms are recipes for computation that occur in every corner of mathematics. A common iterative procedure for projecting population growth reveals how simple orderly events can lead to a variety of behaviors—explosion, decay, repetition, chaos.... Even common elementary school algorithms for arithmetic take on a new dimension when viewed from the perspective of contemporary mathematics: rather than stressing the mastery of specific algorithms—which are now carried out principally by calculators or computers—school mathematics

> can instead emphasize more fundamental attributes of algorithms (e.g., speed, efficiency, sensitivity) that are essential for intelligent use of mathematics in the computer age. Learning to think algorithmically builds contemporary mathematical literacy.

This perspective suggests that students' knowledge of computational algorithms should develop out of problem situations that demand the use of such algorithms. Hence, the power of algorithms resides in their applicability to the process of solving mathematical problems.

Accordingly, the assessment of students' algorithmic knowledge should be aligned with this perspective on mathematics. In particular, rather than presenting students with assessment items directed toward measuring their mastery of algorithmic skills, assessment items should instead focus on problems that foster mathematical reasoning, communication, and connections. The NCTM (1995, p. 11) describes these types of assessments and the role of algorithms in these activities:

> Students engage in solving realistic problems using information and the technological tools available in real life. Moreover, skills, procedural knowledge, and factual knowledge are assessed as part of the doing of mathematics. In fact, these skills are best assessed in the same way they are used, as tools for performing mathematically significant tasks.

Hence, curriculum, instruction, and assessment are aligned with respect to algorithms. The focus of school mathematics is on significant mathematical tasks, and algorithms are a natural, and essential, component of these tasks.

One of the more promising authentic tools for assessing students' mathematical understanding that has emerged from the reform of school mathematics is the performance-assessment task (Romberg 1993). These open-ended tasks—which require students to demonstrate their mathematical power as they "perform" in authentic problem-solving situations—are ideally significant, engaging, rich, active, feasible, and equitable (Stenmark 1991). This demonstration of students' mathematical understanding allows teachers to gain accurate insights into their students' thinking. By building on this information, they can make appropriate instructional decisions for individual students.

STUDENTS' USE OF ALGORITHMS WITHIN A PERFORMANCE-ASSESSMENT TASK

The remainder of this paper examines students' responses to a mathematics performance-assessment task that teachers have created and used to assess their students' mathematical knowledge. In particular, individual students' responses to the task Deer Population are analyzed to reveal the importance of algorithms in the solution processes. The students were tenth graders from

Wisconsin. The examples of responses have been taken from a task developed as part of the Wisconsin Performance Assessment Development Project (WPAD). The WPAD developed performance-assessment tasks for the state of Wisconsin as part of the Wisconsin Student Assessment System. Research and development was conducted at the Wisconsin Center for Education Research under the direction of Norman Webb. This task—which shares the characteristics of quality performance tasks outlined previously—is one of many performance-assessment tasks that have been written, developed, pilot-tested, and scored by mathematics teachers throughout Wisconsin.

A Population Explosion

The task Deer Population (fig. 7.1), which serves as the focus of our discussion, addresses a topic that was important to many students in Wisconsin. Over the past several years the deer population in Wisconsin has been increasing dramatically, and the effects of this growth (e.g., destroying farmers' crops, possibly contaminating the water supply in both urban and rural communities) have received much attention. Several controversial measures have been proposed and enacted to curb this growth. Not surprisingly, these measures have been the focus of extensive media coverage, and the growing deer population has been important to the lives of many students and their families. Consequently, this authentic task, situated in a familiar context, provides students an opportunity for making connections between out-of-school and in-school experiences.

In addition, the task is structured around significant mathematics. In particular, the task focuses on "functions that are constructed as models of real-world problems" and emphasizes "the connections among a problem situation, its model as a function in symbolic form, and the graph of that function" (NCTM 1989, p. 126). The teachers who developed this task chose

Deer Population

Students from Boomer High recently studied a herd of 100 deer living in a nearby forest. Based on the number of female deer they were able to count, they hypothesized that the total deer population could be described by one of the following two functions:

#1: $P(t) = t^2 - t + 100$ or #2: $D(t) = 5t + 100$

t = the number of years after the study.

1. Draw a table or graph of the deer populations represented by each function for every year over a ten year period.

2. Compare and contrast the two functions to describe what they predict will happen to the deer population over an extended period of time.

Fig. 7.1. Deer Population Task

the two functions in order to model a population that was growing linearly and one that was growing exponentially. Both are possible, depending on the nature of the herd, the environmental conditions, and the period of time over which the herd is being studied. The task also requires students to communicate their mathematical knowledge and reasoning in both graphical and written form.

The task is feasible in that students should be able to complete it during class time and most students will be able to respond to the requirements of the task in an appropriate fashion. The task is open ended in the sense that students can choose how they represent the functions and how they go about computing the actual values to represent (e.g., with paper and pencil or a calculator). Moreover, in the second part of the task, students are asked not only to compare and contrast what the two functions are modeling but also to predict the growth of the deer population into the future. This part of the task allows for even more openness in the communication of students' mathematical knowledge and reasoning and takes advantage of the richness of the problem. Finally, the task reveals that algorithms play a significant role in the mathematical-reasoning and problem-solving processes of the students whose responses follow.

Peter's Response

Peter's response (fig. 7.2) demonstrates that he has an understanding of the important features of the problem. In part 1 of the task, he correctly evaluated the total number of deer for each function over the ten-year period and displayed the data by using both tabular and graphical representations. It is interesting to note that Peter did not leave any trace of his work in evaluating the functions. Since calculator use was permitted during the assessment, Peter may have used his calculator to evaluate the number of deer for each successive year (graphing calculators were not allowed). Without a paper trail, however, it is impossible to be sure of the exact algorithmic process Peter used to calculate each of these values.

Peter graphed both functions on the same Cartesian coordinate system and thus displayed the visually salient features needed for comparison. In part 2 of the task, Peter used his graphs of the functions to accurately describe the changes in the deer population that these functions modeled. In particular, he recognized that up to the sixth year, function 2 predicts "a higher growth rate" and that after the sixth year, function 1 predicts "a higher growth rate" for the deer. In describing function 2, Peter revealed a slight misconception. The growth rate of function 2 is actually higher than that of function 1 only until the third year. It is difficult to determine on the basis of Peter's written response if he is actually looking at the rate of growth or if he is incorrectly using this term to describe the cumulative number of deer. To his credit, however, Peter later developed the idea of the rate of growth of the deer population by correctly describing the growth rate of function 2 as a constant five deer each year. Peter also recognized that function 1 increases more rapidly over an extended period of time than does function 2 and furnished insight into the reasoning that led to this recognition when he concluded that "this is because it [function 1] uses the square ... and square numbers grow at [a] much faster rate when they get large."

Peter's response provides evidence that he understands functional notation and can translate the symbolic representation of a function into tabular and

1. Draw a table or graph of the deer populations represented by each function for every year over a ten year period.

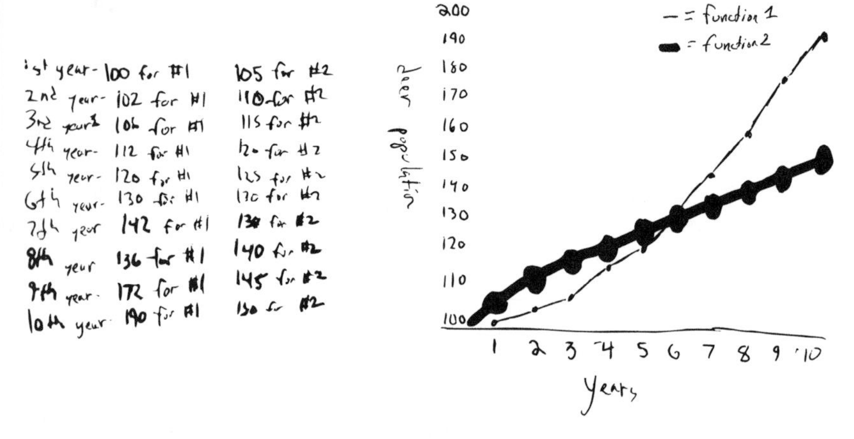

2. Compare and contrast the two functions to describe what they predict will happen to the deer population over an extended period of time.

#2 function predicts a higher growth rate for about six years, after that, it is more conservative than #1. It sets growth rate at a flat 5 deer per year. It lags farther and farther behind #1 as time goes on.

#1 function is slow in the beginning but eventually greatly supersedes #2 in number of deer predicted. This is because it uses the square of the number of years after the study, and square numbers grow at much faster rate when they get large than simple multiplication, which is used in #2.

Fig. 7.2. Peter's response demonstrates that he probably used a calculator algorithmic procedure.

graphical form. Moreover, he was able to use an algorithmic procedure—possibly in the form of substituting the values for each year into the function and procedurally entering the appropriate keystrokes into his calculator—to calculate the values represented in the table and graph. He also recognized the significance of the point of intersection of the two functions' graphs (or the common number of deer at six years for both functions' tabular representations) and provided descriptions of the important features that the graphs display. These descriptions reveal his aptitude for written mathematical communication. Peter also correctly interpreted and described an important distinction between linear and quadratic functions and demonstrated an understanding of the concept of slope as a rate of change. Perhaps most important, Peter's description offers evidence that he was able to make the connection between the functions and the real-world phenomena (the growth of a deer herd) that the functions were modeling.

Susan's Response

Susan's response (fig. 7.3) shows that she understood how to evaluate and graph a function. She correctly evaluated and graphed each function over the ten-year period and unlike Peter, provided a clear indication of the type of algorithm she used to calculate the yearly deer population. She consistently substituted a particular value for the variable *t* and proceeded to perform the calculations that each function dictates. In other words, Susan used a recursive step-by-step computational procedure (an algorithm!) to assist her in solving the problem. On the basis of her work, it appears that, again unlike Peter, she treated the algorithm as a paper-and-pencil computational procedure rather than as a calculator algorithm. However, in calculating the values for function 2, Susan appears to have recognized the pattern of linear growth (increasing by 5) and to have used this pattern rather than the algorithm to calculate the last six values.

It is interesting to note that she evidently treated the evaluation and graphing of each function as two distinct problems. The separation of the yearly deer populations that each function models from their respective graphs (i.e., placing the data representation of function 1 above that of function 2 on two separate Cartesian coordinate systems) makes it more difficult—compared to Peter's organization of the data—to see the features needed to compare and contrast the functions. Consequently, Susan did not provide evidence that she recognized the point at which the two models have equal deer populations.

Susan did correctly indicate that function 2 predicts larger populations at first and function 1 predicts larger populations later, but she neglected to specify the domain for each case, and as mentioned previously, she did not identify the point at which the functions produce equal values. Susan recognized that the graph of function 1 has a "nearly exponential curve" and described the linearity of function 2 as being "more steady" than that of

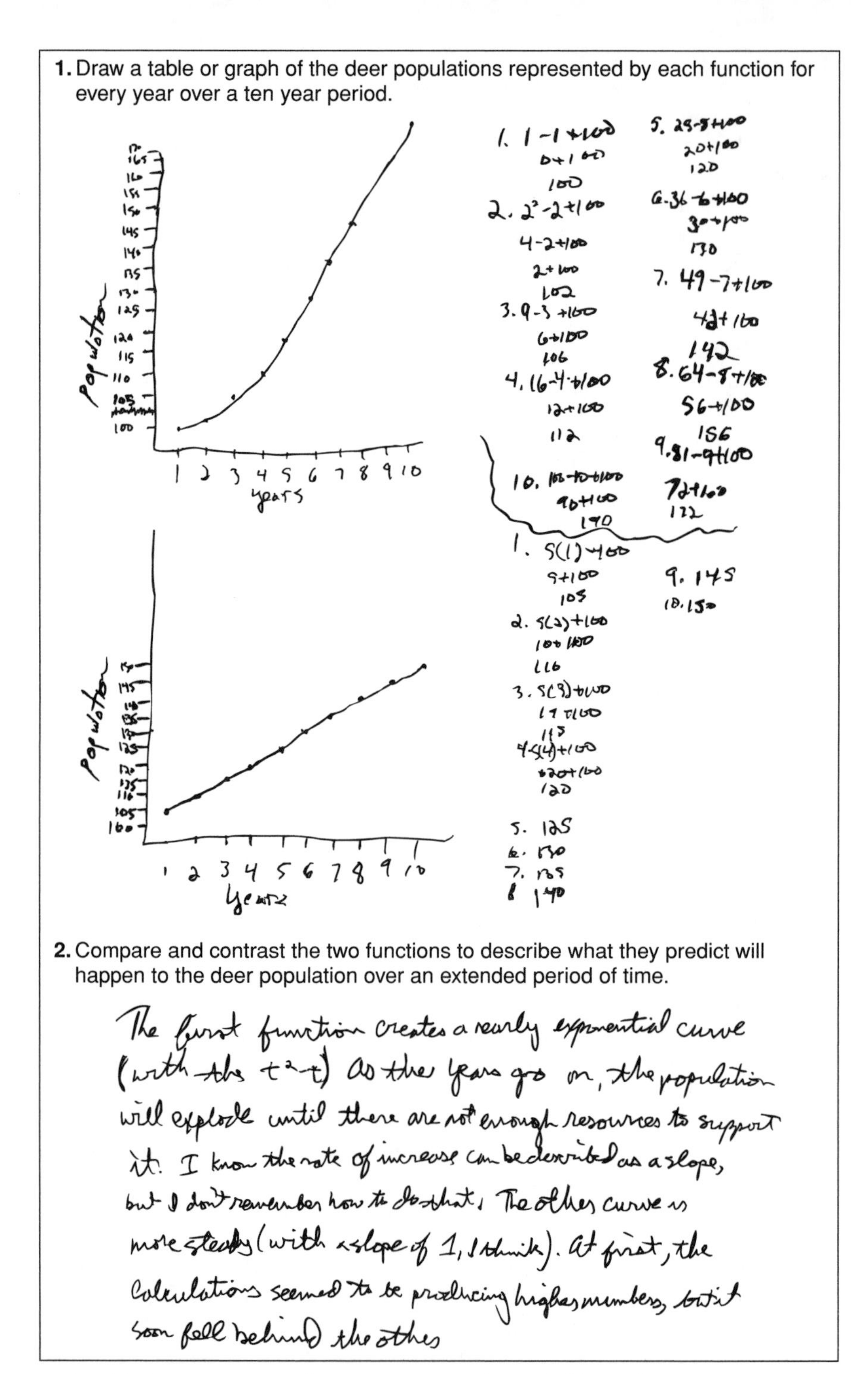

1. Draw a table or graph of the deer populations represented by each function for every year over a ten year period.

1. 1 − 1 + 100
   0 + 100
   100
2. $2^2 - 2 + 100$
   4 − 2 + 100
   2 + 100
   102
3. 9 − 3 + 100
   6 + 100
   106
4. 16 − 4 + 100
   12 + 100
   112
5. 25 − 5 + 100
   20 + 100
   120
6. 36 − 6 + 100
   30 + 100
   130
7. 49 − 7 + 100
   42 + 100
   142
8. 64 − 8 + 100
   56 + 100
   156
9. 81 − 9 + 100
   72 + 100
   172
10. 100 − 10 + 100
    90 + 100
    190

1. 5(1) + 100
   5 + 100
   105
2. 5(2) + 100
   10 + 100
   110
3. 5(3) + 100
   15 + 100
   115
4. 5(4) + 100
   20 + 100
   120
5. 125
6. 130
7. 135
8. 140
9. 145
10. 150

2. Compare and contrast the two functions to describe what they predict will happen to the deer population over an extended period of time.

The first function creates a nearly exponential curve (with the $t^2 - t$) As the years go on, the population will explode until there are not enough resources to support it. I know the rate of increase can be described as a slope, but I don't remember how to do that. The other curve is more steady (with a slope of 1, I think). At first, the calculations seemed to be producing higher numbers, but it soon fell behind the other

Fig. 7.3. Susan's response demonstrates that she probably used a paper-and-pencil algorithmic procedure.

function 1. She also indicated that she understands the concept of slope as a rate of change, but it appears that she did not recall the correct algorithm to calculate the slope of a line and in fact, provided an incorrect slope ($m = 1$ instead of $m = 5$) for the linear model.

Susan's response, like Peter's, furnishes evidence that she understands functional notation and can evaluate individual values of functions and construct their graphs. However, her description overlooks some of the important aspects of the data, which may in part be because of her having graphed the functions separately. These aspects could also have been determined from the function evaluations and might have been more visible had she created a single function table rather than two separate ones. Her description also includes mathematically important concepts such as rate of change and exponential functions; this written account reveals that she is able to communicate her mathematical knowledge and her understanding of the task. Susan's written account also suggests that she has made the connection between the functions and the deer population that they are modeling. Indeed, by discussing the limited resources to support a deer population that she believes is growing exponentially, she perhaps makes a stronger connection to the real world than Peter did.

Kathy's Response

Like Susan, Kathy left clear paper-and-pencil evidence (fig. 7.4) of the algorithmic procedure she used to calculate the yearly deer populations represented by the function. Since Kathy left all her answers as fractions (when appropriate) and showed each step in simplifying the individual values, we might conclude that this algorithm was based on paper-and-pencil calculations rather than on calculator keystrokes. An examination of the details of her algorithmic calculations, however, reveals that she has an incomplete understanding of the symbolic representation of a function. She demonstrated an initial understanding of this type of representation by correctly starting an algorithmic procedure for evaluating the deer populations for each year (e.g., $t = 2$, $D(2) = 5(2) + 100$), and in fact, at the same step for each year in her algorithm, she calculated the correct population (e.g., $t = 2$, $D(2) = 110$). However, it becomes apparent that as Kathy proceeded to solve the resulting equation for D, her algorithm became flawed. It appears that after substituting the appropriate value for t, Kathy mistakenly treated D as the unknown variable rather than as a represention of the function itself. Essentially, she seems to have regarded the functional notation $D(t)$ as denoting the multiplication of the two variables D and t. As a result, she has in a sense combined a part of the algorithm for calculating individual values of a function with the traditional algorithm for solving a single linear equation. Kathy consistently used this significant misconception throughout the problem.

It is difficult to determine solely on the basis of Kathy's response the exact nature of her misconception and how she came to apply this erroneous

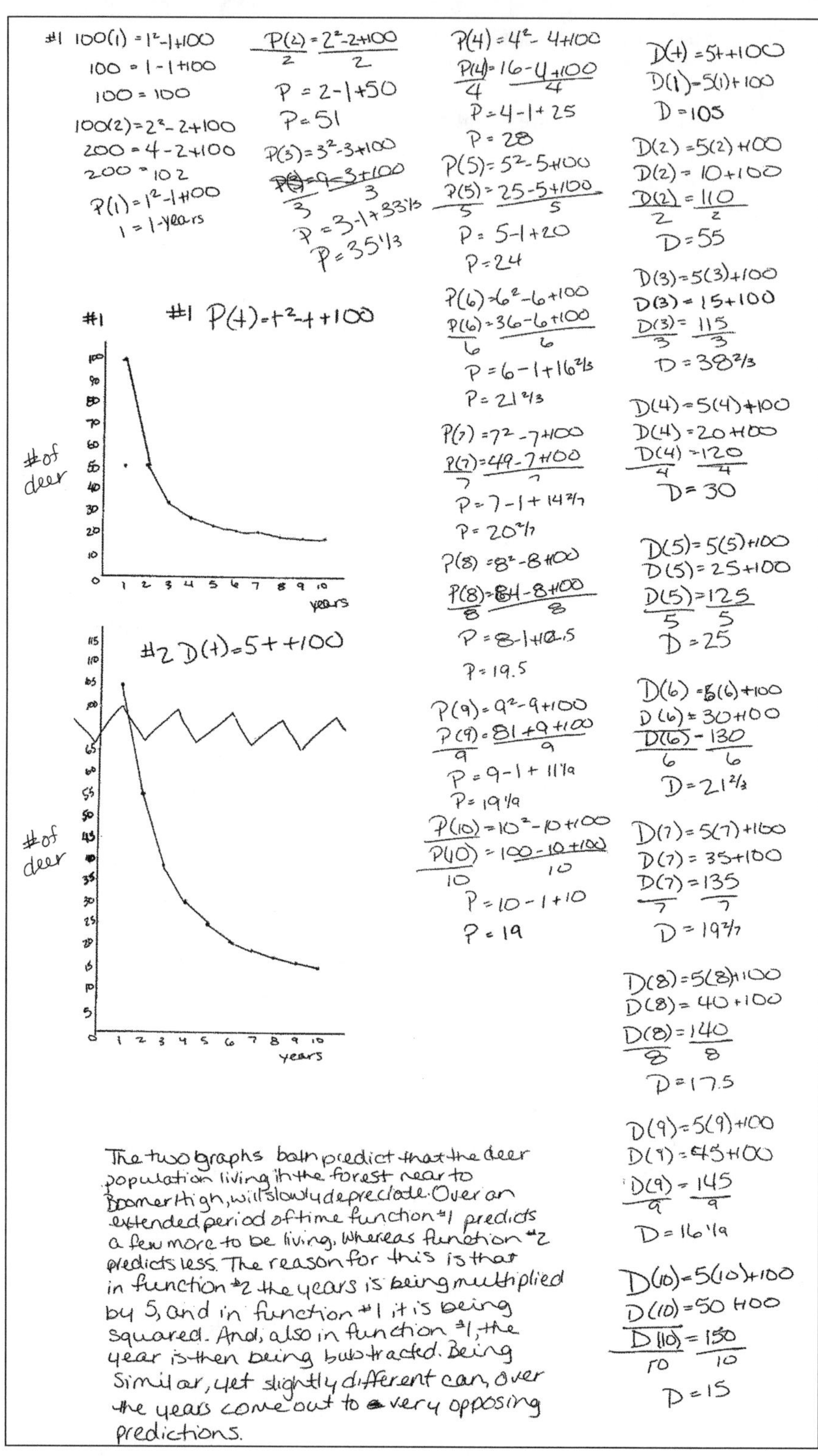

Fig. 7.4. Kathy's response reveals that she used a faulty algorithm.

algorithm. Perhaps she at one time was able to apply both these algorithms, but in responding to this task she could not quite remember the correct algorithm and somehow combined elements of both procedures in an attempt to come up with the individual values of the function. Or it could be that she did not fully understand the algorithm for calculating individual values of a function and this lack of understanding inhibited the correct application of the algorithm in this particular context. Another possible explanation is that she had not encountered this symbolic form of functions and attempted to apply the algorithmic procedures that seemed appropriate to her: substituting particular values for certain variables until she had one equation and one unknown and then applying an algorithm that allowed her to solve for that variable. After further inspection of Kathy's response and on the basis of the initial experimenting she did in trying to substitute 100 for P and particular values for t, we might conclude that the latter explanation seems reasonable. Without further communication with Kathy, however, we are unclear about exactly what her reasoning was in applying "her" algorithm.

It is reasonable to assume on the basis of Kathy's work that she is able to solve linear equations and that she has the knowledge and skills to graph nonlinear functions. Similar to Susan, Kathy has separated the yearly deer populations that each function models from their respective graphs (situating the data representation of function 1 above that of function 2 on two distinct Cartesian coordinate systems). Again, this representation of the data makes it difficult to see the features needed to compare and contrast the two models. Kathy's written description reflects this difficulty in that she does not give specific details in her analysis of the two functions and like Susan, does not provide evidence that she recognizes the point at which the two models have equal deer populations. (It is interesting that even though Kathy has calculated incorrect values, the point of intersection of the two functions determined by her procedure is the same as that determined by using the correct values of the functions: $t = 6$.)

Kathy's description of her displayed data provides an accurate (although not detailed) account of the relationships between the two functions. She noted that the deer populations are declining with time and correctly stated that after "an extended period of time," the population predicted by function 1 will be greater than that predicted by function 2. She did not, however, indicate that she recognizes that function 2 initially models a greater deer population than function 1.

Moreover, Kathy suggested that the reason function 1 predicts a greater population over time than function 2 is because function 1 squares t and function 2 multiplies t by 5. This explanation appears to indicate that Kathy has some understanding of the differences between the rate of change of a quadratic relationship and that of a linear relationship. However, it would also suggest that she has missed an important connection between the symbolic representation of the functions and the graphical and tabular representations. More specifically, she recognized that symbolically, function 1 changes exponentially and function 2 changes linearly, yet she did not realize that her graphical and tabular representations are not consistent with her symbolic interpretation. Hence, she has not made connections among the representations and consequently is not able to verify her work by monitoring the consistency of the representations.

Kathy demonstrated a level of mathematical sophistication comparable in some aspects to that shown by Peter and Susan. Other than one minor error (she calculated $D(3)$ as 38 2/3 rather than 38 1/3), her calculations are precise (although her algorithm is flawed), her graphs are constructed accurately on the basis of her calculations, and her response is well written and contains appropriate mathematical language. Yet she has developed an erroneous understanding of functional notation and demonstrated a faulty algorithm for evaluating the individual values of the functions over a period of time. In addition, even though Kathy's written description provides evidence that she has made the connection between the functions and the deer populations that they model, we could conclude that this connection may not be complete, since her graphical representations indicate that the populations were decreasing, and in reality, the deer population in Wisconsin was increasing.

CONSIDERATIONS

These students' responses reveal some important information about the role of algorithms in the Deer Population task that may have implications for performance assessments in general. The prevalence of algorithms and algorithmic thinking is obvious in all the responses. The students used both calculator-keystroke algorithms and paper-and-pencil algorithms to calculate the values of the functions at different times. The algorithms were characterized by minute differences (reflected in the sequence of arithmetic and algebraic operations) that were unique to the individuals' mathematical knowledge, skills, and experiences. The students' construction of tabular data and their translation of these data into graphical representations—by the plotting of ordered pairs on the Cartesian coordinate system—can also be thought of as an algorithmic procedure. In this sense, the functions themselves can be viewed algorithmically as students rely on the algebraic representation of these functions to determine the functions' behavior. The prevalence of algorithms reinforces the idea that algorithms do indeed arise naturally in problem-solving situations and that they are an important component of mathematics.

It is also important to consider the incorrect algorithm that Kathy used to evaluate the individual values of the functions. Kathy was not the only student who responded to the task by using this type of algorithm. In fact, of the 129 students who attempted to solve this task, 38 students used the same algorithm that Kathy employed (29% of the students). It appears that a number of students have somehow developed the same type of misconception in their process of evaluating this symbolic representation of a function. Interestingly, Herscovics (1989) describes similar results among students responding to items on the National Assessment of Educational Progress. He reports that nearly all students (98%) could correctly evaluate $a + 7$ when $a = 5$ but only 65 percent of these same students could evaluate $f(5)$ when $f(a) = a + 7$. Hence, the introduction of functional notation presents new difficulties for students. Teachers need to be aware of these difficulties and adjust their instruction so that students have a variety of opportunities—in meaningful contexts—to evaluate functions, apply appropriate algorithms, and become familiar with functions of this form.

It is interesting to consider the possibility that students may not be making "substantial" connections among the various representations of the functions. In particular, the students did not provide convincing evidence that they made all the connections among the symbolic representations of the functions, the tabular forms of the functions, the graphical forms of the functions, and the real-world phenomena (the deer population) that all these representations were modeling. This failure to make connections could be a factor of the students' inability to communicate their mathematical reasoning and connections (or their lack of familiarity with doing so). Nevertheless, as teachers we need to give students tasks such as these so that they can develop mathematical connections and communication skills.

A final consideration to keep in mind as we assess students' mathematical knowledge and skills is the notion that their responses reflect what they know and can do rather than what they do not know and cannot do. As teachers we must exercise caution in making inferences about students' knowledge and abilities solely on the basis of their written responses to assessment items. Other means of assessing students' knowledge and abilities, such as observations and interviews, might also be used to get a more complete picture of students' mathematical power. Moreover, if a student does not demonstrate specific knowledge or skills on a task, it does not necessarily follow that the student is deficient in a given area. Several alternative explanations may apply: the student may not be able to communicate the knowledge he or she possesses, the student may not understand the language or context of the task, the task may be poorly constructed, or the student may interpret the task in an unexpected fashion. Whatever the reason, we must always be prepared to ask students to revisit the task and further explain their solutions so that we can achieve a better understanding of their mathematical knowledge. In this way we can make more-informed instructional decisions that are based on our students' mathematical knowledge and abilities.

CONCLUSION

As the reform of school mathematics continues to unfold, algorithms will certainly be a valid and essential component of this reform. Although the emphasis on, and context of, algorithms will not resemble what they have been in the past, the importance of algorithms will not have diminished. On the contrary, algorithms will always provide a cornerstone for a foundation on which students can build their mathematical power.

REFERENCES

Cangelosi, James S. *Teaching Mathematics in Secondary and Middle School: An Interactive Approach.* 2nd ed. Englewood Cliffs, N.J.: Merrill, 1996.

Fowler, D. H. "What Society Means by Mathematics." *Focus* 14 (1994): 12–13.

Herscovics, Nicolas. "Cognitive Obstacles Encountered in the Learning of Algebra." In *Research Issues in the Learning and Teaching of Algebra,* edited by Sigrid Wagner and Carolyn Kieran, pp.60–86. Reston, Va: National Council of Teachers of Mathematics, 1989.

Mathematical Sciences Education Board. *Counting on You: Actions Supporting Mathematics Teaching Standards.* Washington, D.C.: National Academy Press, 1991.

National Council of Teachers of Mathematics. *Curriculum and Evaluation Standards for School Mathematics.* Reston, Va.: National Council of Teachers of Mathematics, 1989.

———. *Assessment Standards for School Mathematics.* Reston, Va.: National Council of Teachers of Mathematics, 1995.

———. *The National Council of Teachers of Mathematics World Wide Web Home Page* (www.nctm.org). Reston, Va.: National Council of Teachers of Mathematics, 1996.

Romberg, Thomas A. "How One Comes to Know: Models and Theories of the Learning of Mathematics." In *Investigations into Assessment in Mathematics Education,* edited by Morgan Niss, pp. 97–111. Dordrecht, Netherlands: Kluwer Academic Publishers, 1993.

Steen, Lynn A. "Pattern." In *On the Shoulders of Giants: New Approaches to Numeracy,* edited by Lynn A. Steen, pp. 1–10. Washington, D.C.: National Academy Press, 1990.

Stenmark, Jean K. *Mathematics Assessment: Myths, Models, Good Questions, and Practical Suggestions.* Reston, Va.: National Council of Teachers of Mathematics, 1991.

U.S. Department of Education. *The Algebra Initiative Colloquium: Volume 2.* Washington, D.C.: U.S. Government Printing Office, 1995.

8

A Brief History of Algorithms in Mathematics

Janet Heine Barnett

Part 2

WHEN asked to define the term *algorithm,* mathematics educators typically offer something along these lines:

> An algorithm is a step-by-step process that guarantees the correct solution to a given problem, provided the steps are executed correctly.

More difficult questions for us to answer are why and how algorithms are important in developing mathematical understanding. What follows is a look at this question through the lenses of three episodes in the history of mathematics in which powerful new algorithms were introduced.

Episode 1: The Hindu-Arabic Numeration System

The need for a system to write numerals is a central concern in all cultures with written records. Methods that have been devised include a simple tally system (Cave Dwellers' Numerals), the Greek cipher system in which each number is represented by a different letter of the alphabet, and the Roman grouping system with its distinctive subtractive feature in which IX represents ten minus one. The three important features of the modern Hindu-Arabic system (place value, the use of 0 as a place holder, and the use of base ten) seem to have been fully developed in India by the eighth century. Variations on the system occurred much earlier, however. In fact, the concept of place value was used by Babylonians as early as the third millennium B.C. The influence of the Babylonian base-sixty numeration system is still seen in trigonometry today, where, for instance, the central angle of a circle contains 6×60, or 360, degrees.

The eventual transmission of the decimal system from India to Western Europe is due to the work of Islamic mathematicians, particularly Muhammad ibn-Musa al-Khwarizmi (ca. 780–850). In his book *Kitab al-jam`wal* *tafriq bi hisab al-Hind (Book on Addition and Subtraction after the Method of the Indians)*, al-Khwarizmi explained how to write any number in the decimal place-value notation. He also demonstrated algorithms based on this notation for a variety of arithmetic operations. The word *algorithm* is, in fact, derived from his name. The term *algebra,* an area in which many algorithms occur, is derived from another of al-Khwarizmi's works, *Al-kitab al-muhtasar fi hisab al-jabr wa-l-muqabala (The Condensed Book on the Calculation of Restoring and Comparing).*

The earliest known use of Hindu-Arabic numerals in Western Europe appeared in the work of Gerbert d'Aurillac (ca. 940–1003), who became Pope Sylvester II in 999. Gerbert represented numbers on a counting board using counters marked with the numerals 1, 2, 3, ... , 9 in columns on the board, with a blank column representing 0. He did not appear to have algorithms for calculating with these counters, however. It was from al-Khwarizmi's works on the Hindu-Arabic system that Leonardo da Pisa (ca. 1180–1250), known today as Fibonacci (son of Bonaccio), and other Western European medieval mathematicians learned these algorithms. Fibonacci's *Liber abaci (Book of Calculation)* used both the notation and the algorithms of the Hindu-Arabic numeration system to solve a wide variety of arithmetic problems of a practical nature. Although it contained little original mathematics, *Liber abaci* did illustrate the effectiveness of the Hindu-Arabic techniques, an aspect of the system that was highly influential with the abacists of fourteenth-century Italy.

The Italian abacists, or *maestri d'abbaco,* were mathematical practitioners who earned their living by teaching methods of computation. Their students were the children of Italian merchants who needed these methods to compete within the commercial setting of Renaissance Italy. Although one might expect the term *abacist* to refer to one who uses an abacus, the Italian abacists actually taught their middle-class pupils the "new" Hindu-Arabic numeration system and its algorithms. This unusual use of the word as a professional title illustrates the extent to which the Hindu-Arabic algorithms had been adopted as superior to the abacus by the Italian practitioners of the time. In fact, since the first Latin translation of al-Khwarizmi's arithmetic text appeared in the twelfth century, a heated battle had raged between those who supported the use of a counting board for computational purposes and the proponents of the Hindu-Arabic system (fig. 8.1).

What led to the eventual adoption of the Hindu-Arabic numerals and their accompanying algorithms was the strong advantages of the system compared to the use of the counting board. The counting-board system was logistically awkward, since it required both a large board and a bag of counters; the sales counters of today's stores are a reminder of the nontransportability of the medieval counting board. More important, recording the computations done with the counting board was difficult, since each step had to be eliminated before the next step could be effected on the board. The

Fig. 8.1. A sixteenth-century publisher's woodcut showing a competition between a supporter of the Hindu-Arabic algorithms and a supporter of the abacus.

Hindu-Arabic algorithms allowed one to record each step of the process, making the verification of the result much easier.

Thus, the practical needs and concerns of the merchant class, the ready availability of paper, and the financial ability of the middle class to educate their children led to widespread acceptance of the Hindu-Arabic numeration system, first in Italy and eventually throughout Europe. Lest it be believed that this acceptance was an easy one, however, we note that arithmetic texts written between the thirteenth and sixteenth centuries used the Hindu–Arabic place-value system to record whole numbers only. Fractions, when needed, were recorded as common fractions (1/2, 3/8) for commercial purposes and Babylonian sexagesimal fractions ($1/2 = 30/60$, $3/8 = 22/60 + 30/60^2$)

for trigonometric purposes. It was not until the sixteenth century that Simon Stevin (1548–1620) combined the idea of a decimal fraction with place-value notation to complete our modern Hindu-Arabic numeration system.

EPISODE 2: SOLVING THE CUBIC EQUATION

Whereas the eventual acceptance of the Hindu-Arabic algorithms for arithmetic was largely due to practical concerns, the sixteenth-century search for an algorithmic solution to the general third- and fourth-degree polynomial equation was anything but a practical matter. Linear equations, however, were of practical interest because of their role in solving problems of proportion, such as the computation of agricultural yields or conversion between monetary units. Methods for solving linear equations appear in texts from ancient China, Babylonia, and Egypt. Both ancient China and Babylonia also had algorithms for solving second-degree (quadratic) equations, which can arise from proportion problems (e.g., $2/x = x/2$) as well as from practical problems involving the Pythagorean theorem. Quadratic-equation algorithms appear in Greek manuscripts as well (albeit in the geometrical form preferred by Greek mathematicians) and eventually found their way to Western Europe through the hands of the Islamic mathematicians who extended and systematized the solution methods. Since negative numbers were not acceptable in these early stages of mathematical development, several different algorithms were needed to handle the various cases resulting from the requirement of positive coefficients. Al-Khwarizmi, for example, presented six types of linear and quadratic equations, showing how each type can be solved algebraically and providing a geometrical justification in each case.

Where, then, does the problem of solving a third-degree (cubic) equation first arise? Babylonian tables giving values of $n^3 + n$ and texts involving problems of the form $a^2n^3 + an = b$ suggest that finding approximate solutions to certain cubic equations was part of the repertoire of Babylonian mathematicians. To date, however, no tables have been found to indicate that the Babylonians possessed an exact (versus approximate) algorithm for solving cubic equations. Attempts to find such an algorithm appear to originate within the realm of pure Greek geometry. Within that realm, problems such as the doubling of a cube using only Euclidean tools led to the study of several curves that might be used to give a "geometrical" solution to a cubic equation. Heirs to the Greek geometry texts, Islamic mathematicians of the tenth and eleventh centuries solved several cubic equations using intersections of different conic sections. The mathematician and poet Omar Khayyám (1048–1131) was the first to systematize these geometric procedures, but he was unable to find algebraic algorithms of the type given by al-Khwarizmi for quadratics.

Finding ourselves back with the abacists of the fourteenth and fifteenth centuries, we see that the Islamic algebraic techniques extended to instances of higher-degree equations. Often, these equations could be solved through a substitution that reduced them back to a quadratic equation. One of the earliest appearances of a nonreducible cubic equation is found in a 1344 work by Maestro Dardi of Pisa. Dardi considers a problem involving compound interest that leads him to the equation $x^3 + 60x^2 + 1200x = 4000$. He then solves this equation using an algorithm that is not generally correct, a fact that he himself acknowledges. Dardi's acknowledgment of this fact is perhaps more interesting than the method itself, since it indicates that the search for an algorithmic solution to the general cubic was becoming a central theoretical concern for mathematicians.

The culmination of this search in the sixteenth century is one of the greatest stories in the history of mathematics. The backdrop of the story lies in the university world of sisteenth-century Italy, a world in which tenure did not exist and appointments were based on a scholar's ability to win public challenges. As a professor at the University of Bologna, Scipione del Ferro (1465–1526) did what anyone might do in these circumstances and guarded as a secret his method for solving cubic equations of the form $x^3 + cx = d$. Before his death, however, del Ferro did disclose his method to his student Antonio Maria Fiore (ca. 1506). Since negative numbers were still not allowed as coefficients of equations, this was only one of the thirteen forms that cubic equations could take. It was enough knowledge, however, to encourage Fiore to challenge Niccolò Tartaglia of Brescia (1499–1557) to a public contest in 1535. Tartaglia, who had been boasting that he could solve cubics of the form $x^3 + bx^2 = d$, accepted the challenge, and went to work on finding a solution for Fiore's form $x^3 + cx = d$. Finding the desired solution just days before the contest, Tartaglia easily won the day, but he declined the prize of thirty banquets to be prepared by the loser for the winner and his friends.

Tartaglia's story does not end there, however. Hearing of the victory, the mathematician and gambler Gierolamo Cardano (1501-1576) wrote to Tartaglia seeking permission to publish the method in an arithmetic book. Cardano eventually did convince Tartaglia to share his method (which he did in the form of a poem), but only under the condition that Cardano would not publish the result (see fig. 8.2). As promised, Cardano did not include the method in his arithmetic text. He did, however, publish a method for solving the cubic in his 1545 work *Ars Magna*, claiming to have found the solution in papers of del Ferro, now twenty years dead. In an unhappy ending, a furious Tartaglia was defeated in a public contest with Cardano's student Lodovico Ferrari (1522–1565), who had discovered a solution to the general fourth-degree (quartic) equation, and the cubic formula discovered by Tartaglia now bears the name Cardano's formula.

When the cube and its things near	**Given**
Add to a new number, discrete,	$x^3 + bx = c,$
Determine two new numbers different *By that one; this feat* *Will be kept as a rule*	**Find** u, v **satisfying both** $u - v = c$
Their product always equal, the same,	**and**
To the cube of a third *Of a number of things named.*	$uv = \left(\frac{b}{3}\right)^3.$
Then, generally speaking, *The remaining amount*	**Then**
Of the cube roots subtracted *Will be your desired count.*	$x = \sqrt[3]{u} - \sqrt[3]{v}.$

Fig. 8.2. Tartaglia's poetic version of Cardano's formula

The story of the search for algorithmic solutions to equations continued almost three hundred years beyond Tartaglia's own story. With an algorithmic solution to equations up to degree 4 in hand, mathematicians turned next to the general fifth-degree (quintic) equation. This problem resisted all efforts at solution until the Norwegian Niels Abel (1802–1829) settled it in a somewhat unexpected way. In a celebrated 1824 pamphlet, Abel proved that a quintic formula (analogous to the well-known quadratic formula involving only basic arithmetic operations and extractions of roots) can never be found. The same is true for equations of higher degree as well, making the search for algorithmic solutions to general polynomial equations hopeless. Then, as often happens in mathematics, Abel's "negative" result produced fruit. Beginning with the central idea of Abel's proof, the French mathematician Evariste Galois (1811–1832) developed the concept of a "group of permutations of the roots of equations" to classify those equations that are solvable by radicals. Today, the concept of a "group" is one of the fundamental structures studied in abstract algebra.

EPISODE 3: THE CALCULUS

Besides its soap-opera overtures and its surprise mathematical ending, the story of the search for algorithmic solutions to polynomial equations is interesting for the innovations in mathematics that were developed along the way. One of the more important of these developments was the introduction and spread of algebraic symbolism. Early abacists, like their immediate and ancient predecessors, employed a prose form of algebra, writing out all problems and solutions in words. The new focus on general methods, as well as the unwieldy nature of prose algebra, led naturally to the introduction of abbreviations and symbols. With the rise in symbolism came insight into the generality of these techniques and eventually an interest in the general structures of the equations themselves. Alongside this extension of Islamic algebra came a renewal of interest in Greek works, especially an interest in the theoretical problems of pure geometry. These two interests came together in the first half of the seventeenth century with the development of analytic, or coordinate, geometry by René Descartes (1596–1650) and Pierre de Fermat (1601–1665).

The idea of representing algebraic equations by geometric curves and vice versa is today so familiar that it is difficult to appreciate the tremendous breakthrough effected by the advent of analytic geometry. Within the Greek tradition of rigor adopted by early Western European mathematicians, the representation of the quantity x^2 or x^3 by a line segment required a radical break with tradition. After all, everyone knew that it was unwise to confuse two-dimensional areas (squares) and three-dimensional solids (cubes) with one-dimensional line segments. Once this barrier was passed, however, it became easy to construct new curves. The new challenge for mathematicians was to find methods for studying the increasingly complicated curves introduced by the techniques of analytic geometry.

Seventeenth-century mathematicians accepted with gusto the challenge of developing techniques for the study of curves. Techniques for computing the areas and arc lengths associated with a curve developed alongside procedures for constructing tangent and normal lines to curves, which related in turn to methods for finding the maximum and minimum values of curves. By the middle of the seventeenth century, mathematicians had solved (often by ingenious arguments) a number of special cases of these problems, which today fall under the purview of calculus. Yet it was not until the work of Isaac Newton (1642–1727) and Gottfried Leibniz (1646–1716) in the latter half of the seventeenth century that calculus as we know it today was born.

What, then, were the contributions of these two men to the study of curves that earned them the title Coinventors of Calculus? First, they recognized the connections among these apparently diverse problems of curves, connections that reduce the entire collection to just two basic operations: *differentiation* and *integration*. Second, they recognized the incredible relation between these two operations, which is expressed in the fundamental theorem of calculus: differentiation and integration are essentially inverse operations, much as subtraction and addition are inverse operations.

To be fair, it should be noted that a statement of the fundamental theorem did appear in the 1670 *Geometrical Lectures* of Isaac Barrow (1630–1677) prior to the work of both Newton and Leibniz. Barrow did not, however, possess the powerful *algorithms* that were the third contribution of Newton and Leibniz to calculus. In regard to problem-solving techniques, it is this last contribution that is the greatest of the three. Prior to this development, mathematicians were forced to develop ad hoc, often cumbersome, geometric techniques for each new curve. The development of general "algebraic" algorithms for differentiation and integration brought with it the ability to solve ever more complicated problems, not only in mathematics but also in the newly developing discipline of physics. The extraordinary *power* of these algorithms is demonstrated by the rapid development of offshoots of calculus in the eighteenth century, including the study of planetary motion, differential equations, differential geometry, infinite series, and complex variables. The extraordinary *efficiency* of these algorithms is demonstrated by the fact that problems that had confounded seventeenth-century mathematicians are now simply exercises in any first-year calculus course. The "infinite trumpet" of Evangelista Torricelli (1608–1647), a three-dimensional solid that has finite volume but infinite surface area, is one of many examples of the source of such problems.

HISTORICAL LESSONS FOR THE CLASSROOM

One of the single most striking common features of the three stories told above is the continuity of the development. At each stage, mathematical activity did not end with the development of the new algorithm, as one might expect. On the contrary, mathematical activity *increased* once the original problem was solved. Not only did each new success give rise to the possibility of yet more success in the future, but the newly developed algorithms also allowed mathematicians to concentrate their creative energy on more complicated problems without having to think about the earlier "steps." This suggests that the search for algorithms is—in a very real sense—the driving force of mathematical development.

It is important to note, however, that the development of algorithms, such as those we find in Hindu-Arabic arithmetic, were not developed explicitly to support the development of later, more sophisticated algorithms, such as those we find in calculus. Rather, the scholars involved in the search for good algorithms were motivated to do so by a problem of mutual interest, sometimes of a practical nature and sometimes of a theoretical nature. This historical observation reinforces what current learning theory suggests for the classroom: algorithms must be presented *in context,* not simply for their own sake. Indeed, if the only thing students are to learn from an algorithm is how to execute it correctly, then why not just give them a technique for fast and efficient algorithms, like a calculator? Clearly, punching the right buttons in the appropriate order leads to a correct solution of a very wide class of problems, making any calculator a powerful algorithmic device in its own right.

In this light, it becomes clear that calculators are not simply a means to bypass the use of algorithms altogether. Recalling the historical episodes discussed earlier, we see that the tremendous algorithmic power of calculators in fact offers great potential for *extending* mathematical understanding. In addition to issues that arise from the technology itself, such as rounding error, estimation, and the order of operations, there are many concepts in our current curriculum that can be reinforced and explored in new ways using computational technology. For example, calculators will perform operations exactly as directed, a feature that reinforces the need for clear notation as a means of communication. The centrality of notation in mathematics is clearly seen in the development of the Hindu-Arabic algorithms, where good notation was a necessary prerequisite to developing efficient algorithms. Although symbolic notation in algebra developed almost after the fact, it did inspire even greater generality once it had been introduced. This increased generality led in turn to the development of analytic geometry, without which calculus is impossible. As in history, developing good notation in order to employ calculator algorithms can both facilitate understanding and serve as the source of inspiration for new problems and algorithms in the classroom.

The use of calculators in the classroom does not, therefore, lead necessarily to decreased mathematical understanding, contrary to the fear of many parents and teachers. Yet the danger that many people sense concerning the use of calculators is very real; it is indeed possible that calculators will not serve our students well in the end. In fact, we have encountered this same difficulty in the past when we allowed the current algorithms we teach to become an end in themselves. Our challenge as educators is to identify *what* is being learned from the algorithm (whether it be a traditional one or not) besides the ability simply to execute it. History shows that relinquishing this tradition of "algorithm for algorithm's sake" will be difficult. Newton himself was concerned that the new algebraic techniques would lead to a loss of geometrical intuition and once wrote that "algebra is the analysis of bunglers in mathematics." Let us hope that our fear of bunglers does not stand in our way.

ADDITIONAL READING

Eves, Howard. *An Introduction to the History of Mathematics with Cultural Connections.* 6th ed. Orlando, Fla.: Saunders College Publishing, 1990.

Katz, Victor J. *A History of Mathematics: An Introduction.* New York: Harper Collins College Publishers, 1993.

Kline, Morris. *Mathematical Thought from Ancient to Modern Times.* New York: Oxford University Press, 1972.

Swetz, Frank J. *Capitalism and Arithmetic: The New Math of the Fifteenth Century.* La Salle, Ill.: Open Court Publishing Co., 1987.

9

Understanding Algorithms from Their History

Barnabas Hughes, O. F. M.

SOMETIME long ago (not too long after A.D. 1200) an unknown person wrote a problem that required algebra for its solution. Instead of solving it, he wrote the phrase *Juxta regulas algorismi,* "according to the rules of al-Khwarizmi." A problem in algebra had been posed, and its solution flowed from an application of one of the rules in a book by al-Khwarizmi, whose book on algebra, written in Persia during the first part of the ninth century, was translated into Latin in the mid-1100s. From its title we get the word *algebra.* The author's name was translated as *algorismus.* From this word arose the English *algorithm,* a word that means a rule or procedure for solving a problem. The choice of the word is a fitting tribute to the man who wrote the first elementary textbook on algebra.

The book contained a complete set of rules for solving standard problems, what we call first- and second-degree equations in one unknown. Translated from the Arabic at least three times, the book was used in schools throughout Europe, either in Latin or in vernacular summaries. The rules always worked. For example, consider the equation *three roots (of a square) and (the number) four equal a square (of a number).* This is al-Khwarizmi's statement but without the words in parentheses. Let us follow his procedure. The left column tells us what to do; the result appears on the right.

1. Take half the number of roots and square it:	$\left(\frac{3}{2}\right)^2 = 2\frac{1}{4}$
2. Add this to the 4:	$4 + 2\frac{1}{4} = 6\frac{1}{4}$
3. Take the square root:	$\sqrt{6\frac{1}{4}} = 2\frac{1}{2}$
4. Add this to half the number of roots:	$\frac{3}{2} + 2\frac{1}{2} = 4$
5. Four is a root.	
6. Its square is 16.	$4^2 = 16$

The general case is, *Roots and numbers equal a square.* (Remember: the *roots* are the roots of a square, the *numbers* are constants, and the *square* is a square number.) To test this, you may want to write another problem that fits the format and follow the steps (apply the rules) to a solution. For example, try *three roots and 40 equal a square; find the value of the square.* Only the positive root appeared in the days of al-Khwarizmi because negative roots were not allowed, although he and his successors could compute with negative numbers.

He had similar rules for the other two cases: *Roots and square equal numbers* and *Roots equal square and numbers.* Modern-day students might be challenged to develop algorithmic rules for solving these two quadratic equations. Notice further the singular noun, *square.* Al-Khwarizmi remarks that if there is more or less than one square, you must make it *one* square by appropriate division or multiplication. Moreover, and very different from the modern way of thinking, the unknown was the value of the square, not its root or x.

The first of two remarkable aspects of this kind of algorithmic thinking is that the problem solver is manipulating *known* numbers. This focus is entirely different from modern algebraic thinking where the emphasis is on the unknown and the manipulation of x or x^2. As long as the student of yesteryear could recognize the type of equation that had to be solved and could remember (or look up) the procedure for solving that type, the value of the unknown was at hand. The second aspect flows from the first: all al-Khwarizmi's algorithms can be used on a hand-held calculator. It's simple: just follow the steps! It may be appropriate here to note that al-Khwarizmi's algorithm is simply that of "completing the square."

This line of thinking, with the emphasis on known numbers, was not a creation of al-Khwarizmi. Whether he knew it or not, he was a direct descendant of the ancient Babylonian mathematicians. They too had algorithms that focused on known numbers. For instance, consider a typical Babylonian exercise (although the word *yard* is modern): *The area of a rectangle is 20 square yards; its perimeter is 18 yards; find its length and width.* The steps of the solution follow.

1. Take half the perimeter and square it: $\left(\frac{18}{2}\right)^2 = 81$

2. Subtract four times the area: $81 - 4(20) = 1$

3. Take the square root: $\sqrt{1} = 1$

4. Add the square root to half of the perimeter: $\left(\frac{18}{2}\right) + 1 = 10$

5. Take half of the sum to get the length: $\frac{10}{2} = 5$

6. Subtract the square root from half of the perimeter: $\left(\frac{18}{2}\right)-1=8$

7. Take half of the difference to get the width: $\left(\frac{8}{2}\right)=4$

To clarify what is happening, let the length and width be x and y and the area be xy. Then follow the steps by appropriately substituting x, y, and xy for the given numbers in the problem. You are working with what is called the fundamental Babylonian identity:

$$(x+y)^2-4xy=(x-y)^2$$

Given two of the three terms, the third is determined. The values of x and y are always found. A challenge for students might be to develop an algorithm that would solve this problem:

> Find the area of a rectangular plot of ground whose perimeter is 32 feet and the difference of its length and width is 4 feet.

Al-Khwarizmi did make one improvement on the algorithmic technique of the Babylonians. He offered geometric proofs for his rules (Read 1969). No evidence has been found that the Babylonians ever proved their rules. Apparently they were satisfied with the fact that the rules always worked.

Algorithms have their place in solving problems, as they have had for thousands of years. Whether we call them rules, procedures, methods, techniques, or algorithms, they need to be explained. History often offers the explanation. Then their utility is apparent, fulfilling, and satisfying.

REFERENCE

Read, Cecil B. "Arabic Algebra, 820–1250." In *Historical Topics for the Mathematics Classroom,* pp. 305–9. Reston, Va.: National Council of Teachers of Mathematics, 1989. Also available in *Historical Topics for the Mathematics Classroom,* Thirty-first Yearbook of the National Council of Teachers of Mathematics, pp. 305–9. Washington, D.C.: National Council of Teachers of Mathematics, 1969.

ADDITIONAL READING

Katz, Victor J. *A History of Mathematics: An Introduction,* pp. 31–35. Chicago: HarperCollins, 1992.

Smith, David Eugene. *History of Mathematics,* Vol. 2. New York: Dover Publications, n.d. (Consult this work for its many algorithms in historical settings.)

10

An Exploration of the Russian Peasant Method of Multiplication

Laura Sgroi

WHEN first presented with an alternative computational algorithm, many people are surprised, since it has never occurred to them that there may be several ways to perform basic operations on numbers. Most people have been taught only one way, so they quite naturally assume that there is only one way. The realization that there are many possible procedures to follow when operating on numbers can change the way that people think of mathematics. They may begin to question their belief that mathematics is a set of preordained rules that must be followed and instead move toward a recognition that mathematics is a way of figuring things out, of making sense of the world. This implies that there are many, many more ways—with many yet to be constructed—to make sense of our world. Believing this, of course, entails a much more exciting and dynamic view of both teaching and learning mathematics.

To begin an analysis of a particular alternative algorithm, in this instance the "Russian peasant" method of multiplication, let us study an example of this algorithm (see fig. 10.1).

~~26~~	~~55~~
13	110
~~6~~	~~220~~
3	440
1	+ 880
	1430

Fig. 10.1

This paper is an expansion and extension of "Mathematics as Exploration," which appeared in the *New York State Mathematics Teachers' Journal*, vol. 43, no. 1 (January 1993): 33–38.

The reader is encouraged to stop reading at this point to identify any patterns or regularities that appear, to try to determine what two factors are being multiplied, and to identify the product.

After some time has been spent studying this example of Russian peasant multiplication, the following discoveries are typically made:

- The number at the bottom of the right-hand column, in this example, 1430, is the product of the two numbers that are at the top of the columns, here 26 and 55.
- The numbers on the right are consistently doubled as one moves down the column.
- The numbers on the left are consistently halved, with any remainders dropped.
- The process of doubling and halving stops when the number on the left equals 1.
- Any right-hand number that is paired with an even left-hand number is discarded, but a right-hand number that is paired with an odd left-hand number is retained.
- The retained numbers on the right are added to arrive at the product.
- This method allows one to multiply without knowledge of the multiplication tables.
- The same product will result if the order of the factors is reversed, although the algorithm will not look the same, as shown in figure 10.2.

55	26
27	52
13	104
~~6~~	~~208~~
3	416
1	+ 832
	1430

Fig. 10.2

Once learners have understood *how* the Russian peasant method of multiplication works, they can be challenged to discover *why* it works, and this investigation is even more interesting. Several questions about this algorithm can be considered: Will this algorithm always work? How do you know? How can you justify this algorithm as a valid procedure? How does this algorithm relate to the standard one involving partial sums? How would you explain this algorithm to someone else?

Answers to these questions do not come quickly. Thinking about these questions is best done over several days. For learners with little experience in solving problems of this type, a suggestion might be made to use the "solve a simpler problem" strategy as in figure 10.3, which shows that $16 \times 15 = 240$.

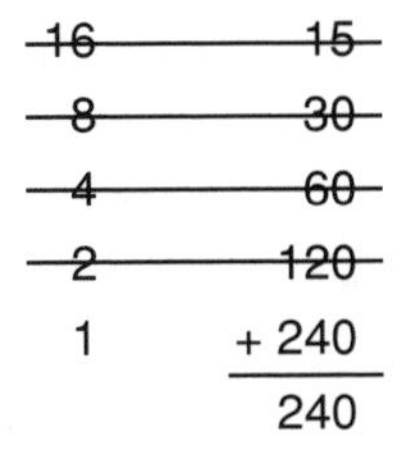

Fig. 10.3

Astute observers will notice that this example is simpler not only because the product is obtained in fewer steps but also because each left-hand number is even, except for the last number. This means that every right-hand number except the last will be crossed out and discarded. With this hint, the reader is again invited to stop reading and take some time to think about *why* this algorithm works.

Learners typically offer at least four different explanations:

1. Some learners opt for a pragmatic explanation, something along the lines of "it works because it works," although their explanation entails more than they realize. These learners interpret 26×55 as twenty-six groups of 55. Each number retained in the right-hand column contributes a certain number of groups of 55 to the final product, as shown in figure 10.4.

~~26~~	~~55~~	~~one group of 55~~
13	110	two groups of 55 (110)
~~6~~	~~220~~	~~four groups of 55~~
3	440	eight groups of 55 (440)
1	+ 880	sixteen groups of 55 (880)
	1430	

Fig. 10.4

These learners notice that the lines in the algorithm that are retained contain two groups, eight groups, and sixteen groups of 55, or twenty-six groups of 55 in all (i.e., 26×55). Subsequent examples of the algorithm with different pairs of numbers always yield a similar result. In each instance of the algorithm, the lines that are retained always result in the correct number of groups of the second factor, and this is always a unique set. *Why* this is so, however, has not yet been answered.

2. If one rewrites the Russian peasant method of multiplying 16×15 in the form of its prime factorization, it would look like this:

$2 \times 2 \times 2 \times 2 \times 1$	$5 \times 3 \times 1$
$2 \times 2 \times 2 \times 1$	$5 \times 3 \times 2 \times 1$
$2 \times 2 \times 1$	$5 \times 3 \times 2 \times 2 \times 1$
2×1	$5 \times 3 \times 2 \times 2 \times 2 \times 1$
1	$5 \times 3 \times 2 \times 2 \times 2 \times 2 \times 1$

This factorization makes explicit what actually happens when the left side is halved and the right side is doubled. In each subsequent step, a factor of 2 is "moved" from the left side to the right side. The compensation process ends when there are no factors other than 1 on the left side.

Most learners find this explanation understandable and interesting, but it does not clearly show why the remainder is dropped from the left-hand number when the halving results in a remainder.

3. Some learners who understand the compensatory nature of the algorithm investigate what would happen if the halving was an accurate halving, that is, with no remainders being dropped. This version of the algorithm is shown in figure 10.5.

26	55
13	110
6.5	220
3.25	440
1.625	880

Fig. 10.5

Decimals are introduced when the number on the left does not halve evenly—that is, when the number on the left is odd. Introducing decimals into the algorithm seems to have two disadvantages. First, the algorithm becomes much more complex, and the halving and doubling is no longer easily done mentally. Second, the algorithm never really ends, since the left-hand column never equals 1 (unless the first factor is a power of 2, as in 16×15).

In the original version of 26×55, the compensation was not always a true compensation, since a certain part of the 55 was, in a sense, "lost." That is why the rows having odd numbers in the left-hand column are retained and those having even numbers in the left-hand column discarded. In this way, no part of the 55 is "lost," and the compensation is a true compensation.

This insight also explains an aspect of the algorithm that many people find puzzling. When one multiplies horizontally across a row of the algorithm (multiply the factor in the left-hand column with its corresponding factor in the right-hand column), sometimes the product is the final product of the algorithm and sometimes it isn't. Realizing that some rows in the algorithm represent true compensations and some do not indicates that prior to the dropping of any remainders, multiplying horizontally across a row will result in the final product. However, once a remainder has been dropped, the compensation is no longer a true compensation. This is what necessitates the retention of certain rows—the rows with an odd number in the left-hand column.

4. Some learners notice that the right-hand column of the algorithm contains one, then two, then four, then eight, then sixteen groups of the second factor and wonder if there is a connection between the Russian peasant algorithm and the binary system. They may hypothesize that there is a unique collection of the right-hand factors that always results in precisely the necessary number of groups of the second factor, since there is a unique way to represent any number in base two.

For example, 26 written in base two is 11010. This collection of digits, 1-1-0-1-0, corresponds to the eveness or oddness of the left-hand number in each row in the algorithm. Reading from the bottom up, a "1" signifies an odd row and a "0" signifies an even row. Every row that corresponds to a 1 in the binary system is odd and will be retained, and every row that corresponds to a 0 in the binary system is even and will be discarded. Hence, just as 26 equals 16 + 8 + 2, so twenty-six groups of 55 equals sixteen groups plus eight groups plus two groups of 55 (see fig. 10.4).

Up to now, this analysis of an alternative algorithm has engaged learners in the following topics: alternative methods of computation, identifying patterns, solving a simpler problem, hypothesizing and testing hypotheses, and predicting. This investigation has encouraged a certain creative approach to solving a problem and presented an opportunity to construct explanations. If people believe that mathematics is a creative endeavor as opposed to a fixed body of knowledge, they may believe that teaching and learning mathematics is a process of exploration and construction as opposed to an accumulation of skills. If teachers and learners can be persuaded that the discipline of mathematics is still growing and evolving, perhaps they can also be persuaded that their teaching and learning of mathematics needs to continually grow and evolve.

BIBLIOGRAPHY

National Council of Teachers of Mathematics. *Historical Topics for the Mathematics Classroom,* Thirty-first Yearbook of the National Council of Teachers of Mathematics. Washington, D.C.: National Council of Teachers of Mathematics, 1969.

———. *Projects to Enrich School Mathematics, Level 1.* Edited by Judith Trowell. Reston, Va.: National Council of Teachers of Mathematics, 1990.

11

Hammurabi's Calculator

Clifford Wagner

THERE is a more-than-two-thousand-year-old algorithm that finds square roots with excellent accuracy. This algorithm, often called the Babylonian method, may have been the basis for an approximation of $\sqrt{2}$ that appears on a Babylonian clay tablet from around the time of King Hammurabi (ca. 1792–50 B.C.) or even earlier (Bailey 1989; Neugebauer and Sachs 1945, pp. 1, 42–44; Struik 1987, pp. 26–31). The Babylonian approximation for $\sqrt{2}$ has an error of less than one-millionth!

The Greek mathematician Heron of Alexandria published this algorithm in the first century A.D., and thus it is also known as Heron's method. In fact, the Babylonians did not actually write a description of their method (Neugebauer 1957, p. 50), and some speculate that they simply used trial and error to find square roots (Powell 1995), so Heron appears to have a legitimate claim for this method. Nevertheless, the algorithm is widely called the Babylonian method and may actually be four thousand years old.

This algorithm is good to learn because it is useful, simple, accurate, and efficient. Moreover, the Babylonian method shows us the marvelous accomplishments of ancient mathematicians, and it nicely illustrates the concepts of iteration and convergence. In addition, the Babylonian method is a special case of the Newton-Raphson method, which is widely known and used today.

A discussion of one or more aspects of the Babylonian method can add a relevant historical illustration to any of the following secondary school mathematics topics: decimal approximations of an irrational number, Newton's method for finding roots, and mathematics in other cultures.

How Does It Work?

Suppose that N is a positive real number whose square root is irrational. The obvious approach to finding the square root of N is to find a rational number that approximates $\sqrt{N}$. As a first approximation, let x_0 be the least whole number for which $N < x_0^2$. Requiring that x_0 be an integer greater than $\sqrt{N}$ simplifies our discussion, but the method also works if x_0 is any rational number near $\sqrt{N}$. Now x_0 approximates $\sqrt{N}$ and the error of this initial approximation is less than 1 because

$$(x_0 - 1)^2 < N < x_0^2$$

implies that

$$x_0 - 1 < \sqrt{N} < x_0,$$

which in turn implies that

$$\left| x_0 - \sqrt{N} \right| < 1.$$

An important attribute of the Babylonian method is that it is iterative; the current approximation can be improved by calculating a new approximation based on the formula

$$x_{n+1} = \frac{1}{2}\left(x_n + \frac{N}{x_n} \right),$$

where $n = 0, 1, 2, \ldots$. For example, the sequence of rational approximations to $\sqrt{2}$ begins with the three approximations

$$x_0 = 2,$$

$$x_1 = \frac{1}{2}\left(2 + \frac{2}{2} \right) = \frac{3}{2} = 1.5,$$

and

$$x_2 = \frac{1}{2}\left(\frac{3}{2} + \frac{2}{\frac{3}{2}} \right) = \frac{17}{12} = 1.41\overline{6}.$$

WHY DOES IT WORK?

To gain an intuitive understanding of why the Babylonian algorithm works, recognize that if x_n overestimates $\sqrt{N}$, then N/x_n underestimates $\sqrt{N}$, and the average of these two estimates,

$$x_{n+1} = \frac{1}{2}\left(x_n + \frac{N}{x_n} \right),$$

is an improved estimate of $\sqrt{N}$.

To understand the algorithm graphically, first consider figure 11.1, which shows both the line $y = x$ and the iteration function $f(x)$ on which the Babylonian algorithm is based when $N = 2$. The line and function intersect at $\left(\sqrt{2},\sqrt{2}\right)$.

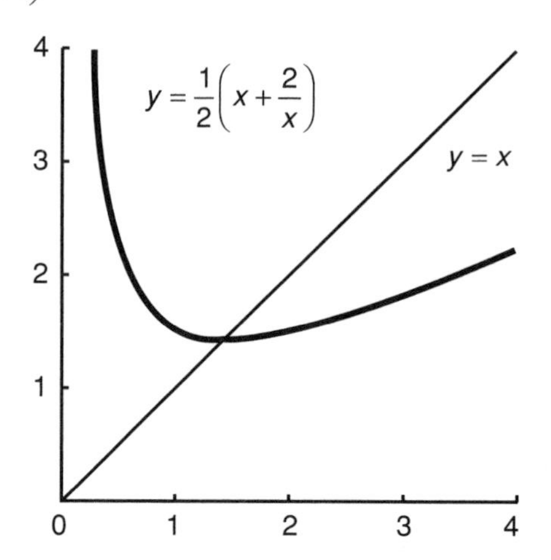

Fig. 11.1. The graphs of $y = x$ and the function
$$y = f(x) = \frac{1}{2}\left(x + \frac{2}{x}\right)$$

Figure 11.2 shows an enlargement of these graphs in a region about the intersection point. The progress of the Babylonian algorithm can be traced by following the cobweb graph, beginning on the horizontal axis at x_0 and following alternating vertical and horizontal line segments, ending a vertical segment when it meets the function $f(x)$, and ending a horizontal segment when it meets the line $y = x$. Wagner (1982) offers a more detailed discussion of cobweb graphs, Newton's method, and other iterative methods.

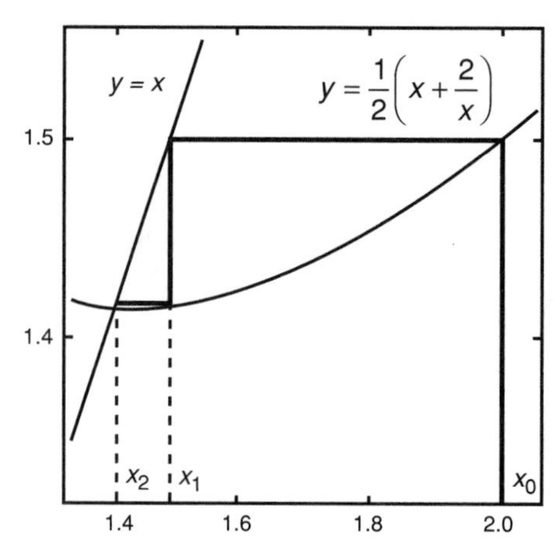

Fig. 11.2. The convergence of $x_0, x_1, x_2, \ldots$ to $\sqrt{2}$

HOW WELL DOES IT WORK?

First let us establish that all the iterates $x_0, x_1, x_2, \ldots$ are greater than $\sqrt{N}$. This can be proved algebraically, since for $x_n > 0$, the inequality

$$\frac{1}{2}\left(x_n + \frac{N}{x_n}\right) > \sqrt{N}$$

is equivalent to

$$\left(x_n - \sqrt{N}\right)^2 > 0,$$

and thus $x_n > \sqrt{N}$ must imply $x_{n+1} > \sqrt{N}$. A rigorous proof that all iterates are greater than the square root of N can be based on the principle of mathematical induction. A secondary school discussion could just as well be based on a computer demonstration, either by calculating several iterations and observing that the approximations are always a bit too large or by drawing a graph such as that in figure 11.1 and observing that all points on the graph of the iteration function are on or above the line $y = \sqrt{N}$ and thus all values of this function (i.e., all approximations) are greater than $\sqrt{N}$.

Now, the error of an approximation produced by the Babylonian method is defined by

$$\varepsilon_n = \left|x_n - \sqrt{N}\right|$$

where $n = 0, 1, 2, \ldots$. Since each x_n is greater than $\sqrt{N}$, each error is simply

$$\varepsilon_n = x_n - \sqrt{N}.$$

Consequently, twice the error ε_{n+1} is

$$\begin{aligned}
2\varepsilon_{n+1} &= 2x_{n+1} - 2\sqrt{N} \\
&= x_n + \frac{N}{x_n} - 2\sqrt{N} \\
&= \frac{x_n^2 - 2x_n\sqrt{N} + N}{x_n} \\
&= \frac{\left(x_n - \sqrt{N}\right)^2}{x_n} \\
&= \frac{\varepsilon_n^2}{x_n}.
\end{aligned}$$

It follows that

$$\varepsilon_{n+1} = \frac{1}{2} \cdot \frac{\varepsilon_n^2}{x_n}.$$

And since $x_n > \sqrt{N}$, it then follows that

$$\varepsilon_{n+1} < \frac{1}{2} \cdot \frac{\varepsilon_n^2}{\sqrt{N}}.$$

When $N > 1$, these error bounds assure us that each subsequent error is less than one-half the square of the previous error. For example, the errors of the first three approximations of $\sqrt{2}$ are bounded by the inequalities $\varepsilon_0 < 1$, $\varepsilon_1 < 0.5$, and $\varepsilon_2 < 0.125$, respectively. Note that these inequalities give upper bounds on the errors; the actual errors are less (and possibly much less) than the calculated bounds.

PROBLEMS FOR FURTHER EXPLORATION AND DISCUSSION

1. How many iterations of the Babylonian method are needed to approximate $\sqrt{2}$ with (*a*) an error of less than one-millionth (the accuracy achieved by Hammurabi's scribes) or (*b*) the same accuracy as your electronic calculator?
2. Many people remember $\sqrt{3}$ as 1.732 because George Washington was born in 1732. Verify this approximation using the Babylonian method. How many iterations are needed to reach an approximation of 1.732?
3. Using the Babylonian method, calculate the first three approximations for $\sqrt{1.44}$. Find x_0, x_1, x_2, and the corresponding errors ε_0, ε_1, ε_2.

REFERENCES

Bailey, D. F. "A Historical Survey of Solution by Functional Iteration." *Mathematics Magazine* 62 (June 1989): 155–66.

Neugebauer, Otto. *The Exact Sciences in Antiquity.* 2nd ed. Providence, R.I.: Brown University Press, 1957.

Neugebauer, Otto, and Abraham Sachs. *Mathematical Cuneiform Texts.* New Haven, Conn.: American Oriental Society, 1945.

Powell, Marvin A. "Metrology and Mathematics in Ancient Mesopotamia." In *Civilizations of the Ancient Near East,* vol. 3, edited by Jack M. Sasson, pp. 1941–57. New York: Charles Scribner's Sons, 1995.

Struik, Dirk J. *A Concise History of Mathematics.* 4th rev. ed. New York: Dover Publications, 1987.

Wagner, Clifford H. "A Generic Approach to Iterative Methods." *Mathematics Magazine* 55 (November 1982): 259–73.

12

Capsule Lessons in Alternative Algorithms for the Classroom

Diane E. Mason

"WE'VE *always* done it this way!" is a common response to suggestions for change. But mathematics has not always been done in the same way. Although the calculations performed centuries ago are still valid today, the algorithms used most likely are not the same. As our number system became more defined and new notations were created, mathematicians developed new algorithms. Yet the historical algorithms illustrate thought. Comparing historical algorithms with those of today can show students how the human understanding of number has developed through time.

Following are three capsules that illustrate alternative algorithms from the pages of history. The first algorithm is a method used for many years to calculate sums and differences. The basic concept behind it is found in the ancient cultures of Japan, Rome, and many European countries. The second algorithm described here, casting out nines, applied modular arithmetic before it was known as such. It was used as early as the ninth century and also claims many cultures, from Middle Eastern to European, in its history. A very different algorithm for approximating square roots is illustrated in the third capsule. Students may find it interesting to compare the accuracy of the values yielded by that method with that of the values given by their calculators.

The study of these historical algorithms will make our students aware that mathematics has a history. The evolution of mathematics is made evident by the study of these and other historical algorithms. The hope is that students will ask, "Why did we change?" "Who had these ideas?" "Why does it work?" Investigating these questions is motivating for students and shows that mathematics is congenial, and the investigation is also mathematics!

CAPSULE 1: RECKONING ON LINES

Reckoning on lines is an algorithm for calculating sums and differences that dates to fifteenth-century Europe. It is a modification of the Japanese abacus and the Roman calculating board. This method of reckoning was used by merchants to total the prices of the purchases made by their customers. Shops contained a table with parallel lines etched on the top. The merchant manipulated small counters on these lines and on the spaces between the lines to represent numerical values (see fig. 12.1). (Hence, we still pay for our purchases at the front "counter" in a store.) The procedure is described by Robert Recorde in his 1542 book *Ground of Artes* (see fig. 12.2).

Fig. 12.1. Counter reckoning in 1514 from Smith (1958, p. 182). Reprinted with permission.

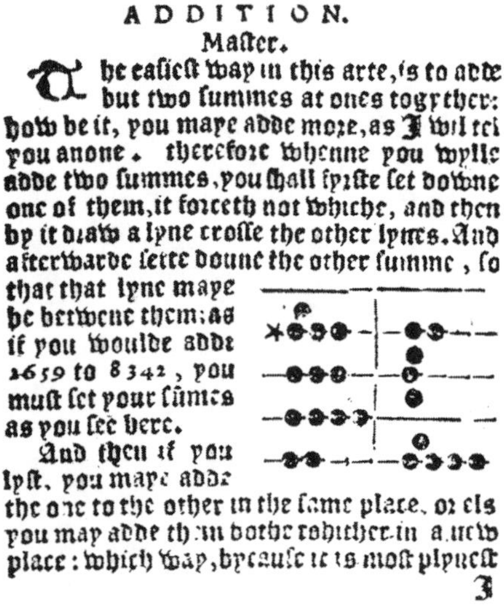

ADDITION.

Master.

The easiest way in this arte, is to adde but two summes at ones togyther: how be it, you maye adde more, as I wil tel you anone. therefore whenne you wylle adde two summes, you shall fyrste set downe one of them, it forceth not whiche, and then by it draw a lyne crosse the other lynes. And afterwarde sette doune the other summe, so that that lyne maye be betwene them; as if you woulde adde 1659 to 8342, you must set your summes as you see here.

And then if you lyst, you maye adde the one to the other in the same place, or els you may adde them bothe togither in a new place: which way, bycause it is most playnest

I

Fig. 12.2. Recorde's description of reckoning on lines from Smith (1958, p. 182). Reprinted with permission.

Procedure

First, draw a set of horizontal parallel lines, then represent the first addend on the lines and in the spaces according to the rules in figure 12.3. One can draw as many lines as required by the numbers in the calculation, with each line representing ten times the value of the line immediately below it.

After you have represented the first number, draw to the right of the counters a line perpendicular to the original lines, and represent the second number to the right of the perpendicular line. If more than two numbers are to be added, continue to append the additional addends to the right as needed and separate the addends with additional lines.

Next, combine the counters by removing the lines separating the numbers. Whenever five counters appear on a line, remove them and exchange them for one counter placed in the space above the line. Whenever two counters appear in a space, remove them and exchange them for a counter placed on the line above. Repeat this process until rule 2 is once again satisfied. Then use the number of counters and the place values to interpret the sum according to rule 1.

Advantages

The addition is actually accomplished by the exchanging process. An advantage of this algorithm is that addition can be done by students who have not mastered the addition facts. True, using memorized facts is faster, but manipulating counters reinforces the concepts of place value and regrouping (by actually *carrying* the counters) and the underlying concept of addition as the union

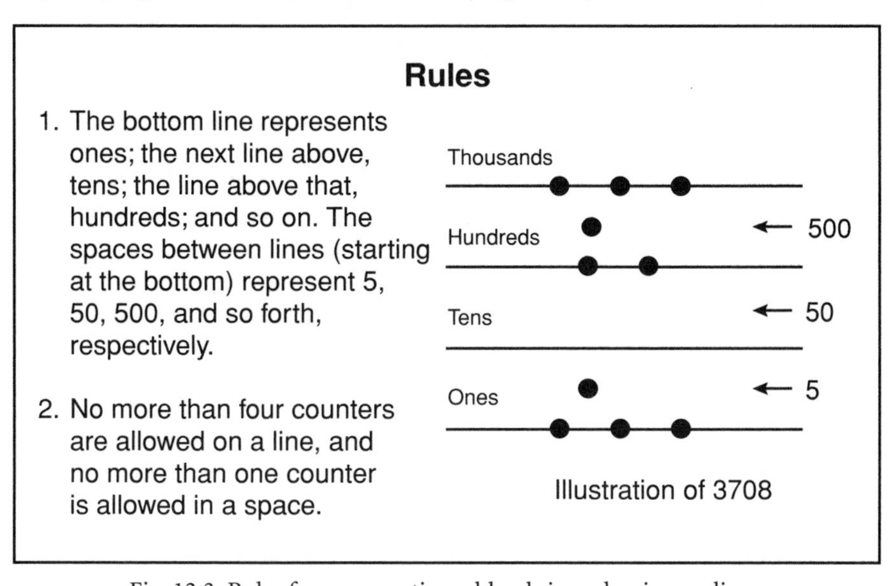

Fig. 12.3. Rules for representing addends in reckoning on lines

of two disjoint sets. The manipulative process is especially helpful for remedial students, since they can both see and feel the addition.

The exchanging of counters is analogous to trading five pennies for a nickel, two nickels for a dime, five dimes for a half-dollar, and so on. Thus, this algorithm allows students to practice monetary exchange as well.

Extension

Once students can add on the counting table, it is not difficult to show them how to subtract. Subtracting on the table also emphasizes the concept of regrouping, since the students must exchange counters when not enough are available on a line or a space.

To subtract, first represent the minuend on the parallel lines. To the right of a perpendicular line, represent the subtrahend. Now remove from both sides of the perpendicular line equal numbers of counters in each position until the right side is void of counters. The counters remaining (to the left of the perpendicular line) represent the difference.

If you need to remove more counters from a line than you have on the line, exchange one counter in the space above for five counters to place on the line in question. If there is no counter in the space, exchange one counter on the line above for two in the space, then take one of those two to exchange for five counters on the original line. Borrowing in this manner reinforces the concept of trading equals for equals (e.g., 1 ten for 2 fives, etc.).

Students who have worked with this algorithm have enjoyed it; it makes sense of the operations. Students may also find that they are more proficient with pencil-and-paper algorithms after exploring reckoning on lines.

CAPSULE 2: CASTING OUT NINES, THE "BOOKKEEPER'S CHECK"

The method of checking calculations known as casting out nines has been known for centuries and has an Indian or Arabic origin (Al-Daffa 1977, p. 95). The technique became very popular in Europe, where it was also known as the bookkeeper's check. In Europe, sometimes numbers other than nine were also cast out for purposes of checking (Smith 1958). The method can be an application of modular arithmetic (Lauber 1990; Smith 1958).

Procedure

To cast out nines from a given number, divide the given number by 9, then discard the quotient and record the remainder. A preferred shortcut for obtaining the remainder is to add the digits of the given number and subtract (that is, cast out) 9 every time the sum is greater than 9. For example, let 83 128 be the given number. To cast out nines by adding the digits, let us work from left to right. $8 + 3 = 11$; $11 > 9$, so cast out nine: $11 - 9 = 2$. Add the 2 to the next digit in the given number: $2 + 1 = 3$; $3 < 9$, so continue adding. $3 + 2 = 5$; $5 < 9$, so continue adding. $5 + 8 = 13$; $13 > 9$, so cast out nine: $13 - 9 = 4$. This process is equivalent to dividing 83 128 by 9 and taking the remainder. Alternatively, simply add the digits of the given number ($8 + 3 + 1 + 2 + 8$), and continue adding the digits of each sum until the sum is a single digit. Thus $8 + 3 + 1 + 2 + 8 = 22$; $2 + 2 = 4$.

To use casting out nines to verify calculations in multiplication, cast out nines from the multiplicand and multiplier. In the example 3921×436, casting out nines yields 6×4. Multiply the resulting numbers, and cast out nines from the product and from the product of the original calculation:

$$
\begin{array}{rcl}
3921 & \equiv & 6 \pmod 9 \\
\times \quad 436 & \equiv & \times\ 4 \pmod 9 \\
\hline
18426 & & 24 \equiv 6 \pmod 9 \\
9863 & & \\
12684 & & \\
\hline
1385456 & \equiv & 5 \pmod 9
\end{array}
$$

Since casting out nines from 24 yields 6 but casting out nines from 1 385 456 yields 5, we are alerted that an error has been made in the computation. After careful inspection we find that the partial products have not been calculated correctly. Once the work is corrected and the actual product—1 709 556—is obtained, we check the work again by casting out nines:

$$
\begin{array}{rcl}
3921 & \equiv & 6 \pmod 9 \\
\times \quad 436 & \equiv & \times\ 4 \pmod 9 \\
\hline
23526 & & 24 \equiv 6 \pmod 9 \\
11763 & & \\
15684 & & \\
\hline
1709556 & \equiv & 6 \pmod 9
\end{array}
$$

Advantages

Casting out nines is an easy procedure that catches most errors. For this reason, it was very popular in Europe before the days of electronic calculators. Although casting out nines does not always ensure that a calculation is correct, we can be certain that an error has been made in computation if a check that has been properly carried out does not agree with the original calculation.

Another advantage to casting out nines is that it can be used to check calculations with all operations, although the application of the algorithm to division may become complicated. The following examples demonstrate the method used to check subtraction and division:

$$
\begin{array}{rcl}
2147 & \equiv & 5 \pmod 9 \\
-\ 1588 & \equiv & 4 \pmod 9 \\
\hline
559 & \equiv & 1 \pmod 9
\end{array}
$$

```
     255R11
23)5876
   46
   127
   115
    126
    115
     11
```

Check:

$$23 \times 255 + 11 = 5876 \equiv 8 \pmod 9$$

$$5 \pmod 9 \times 3 \pmod 9 + 2 \pmod 9 = 8 \pmod 9$$

Students need to be reminded often to check their work. Casting out nines presents another opportunity to remind them of this important step. It also gives them another tool to use in the checking process.

CAPSULE 3: APPROXIMATING ROOTS

Aside from being an interesting curiosity from Greek number theory, figurate numbers are useful to describe the concepts of squaring and cubing and taking square and cube roots. The Greeks represented numbers with arrays of dots. The number of dots implied the number. If the number of dots can produce a square, the number is known as a *square number.*

A Procedure for Finding Roots with Figurate Numbers

To the Greeks, the square root of a number was the number of dots along one side of the square that represents the number. The idea was also illustrated by the drawing of *gnomons,* which when added to a given configuration, enlarge that configuration but allow it to retain its original shape (see fig. 12.4). For instance, if we start with a square of two dots by two dots, we would need to add to that square three dots across the bottom and two more along the side to make a square of three dots by three dots. Similarly, we could enlarge this 3 × 3 square to a 4 × 4 square. This L-shaped addition is a gnomon for increasing a square. The number of complete gnomons is the square root of the number. It is easy to see the one-to-one correspondence between the number of gnomons and the number of dots along one side of the square.

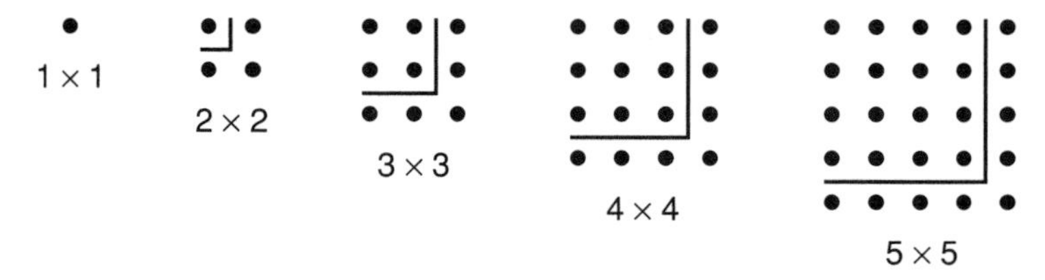

Fig. 12.4. Adding gnomons to an existing square produces successively larger squares.

It is fascinating to notice how closely the roots of nonsquare numbers could be approximated by using the Greek method of gnomons. Arrange the appropriate number of dots in an array that is as nearly square as possible, then draw the gnomons. The number of complete gnomons is the whole-number portion of the square root. Next, fill in the figure with *x*s to complete the last gnomon. Tally the number of dots in the last gnomon and divide the result by the total of the marks (dots and *x*s) in the last gnomon. Add this fraction to the whole number. The result is, of course, a rational number but forms a very reasonable approximation of the given irrational number.

Consider the examples in figure 12.5.

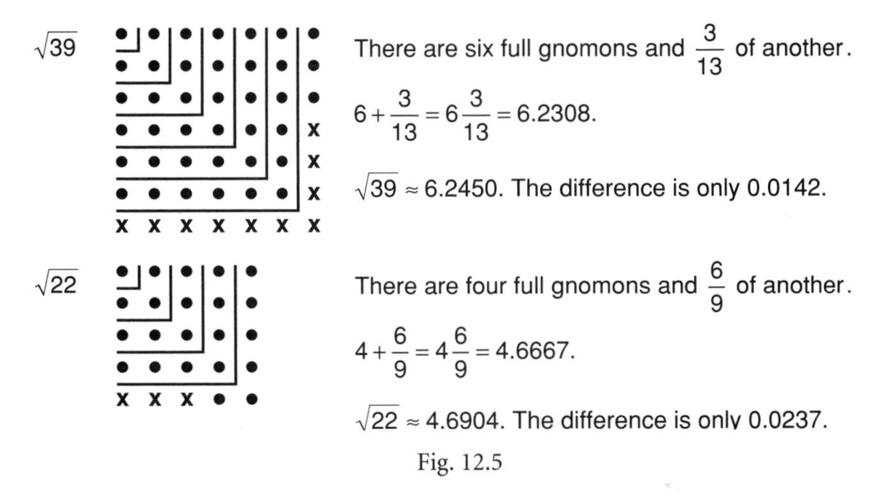

Fig. 12.5

Extension

The same concept is easily extended to three dimensions by using sugar cubes or linking cubes. Cube roots are not as easy to see, since the cube must be disassembled to find the three-dimensional gnomons.

Advantages

Using figurate numbers to explore squares, cubes, and their roots helps the students visualize these abstract concepts. Furthermore, a discussion of methods of approximation would be a beneficial aside to this exploration.

REFERENCES

Al-Daffa, Ali Abdullah. *The Muslim Contribution to Mathematics.* Atlantic Highlands, N.J.: Humanities Press, 1977.

Lauber, Murray. "Casting Out Nines: An Explanation and Extensions." *Mathematics Teacher* 83 (November 1990): 661–65.

Smith, David Eugene. *History of Mathematics.* Vol. 2. 1925. Reprint, New York: Dover Publications, 1958

ADDITIONAL READING

Capsule 1

Freitag, Herta Taussig, and Arthur H. Freitag. *The Number Story.* Washington, D.C.: National Council of Teachers of Mathematics, 1960.

Johnson, Donovan A., and William H. Glenn. *Computing Devices.* St. Louis: Webster Publishing Co., 1961.

Capsule 2

Glenn, William H., and Donovan A. Johnson. *Shortcuts in Computing.* St. Louis: Webster Publishing Co., 1961.

Wheeler, Mary L. "Check-Digit Schemes." *Mathematics Teacher* 87 (April 1994): 228–30.

Capsule 3

Eves, Howard Whitley. *An Introduction to the History of Mathematics.* 5th ed. Philadelphia: Saunders College Publishing, 1983.

Freitag, Herta Taussig, and Arthur H. Freitag. *The Number Story.* Washington, D.C.: National Council of Teachers of Mathematics, 1960.

13

Historical Algorithms
Sources for Student Projects

Rheta N. Rubenstein

STUDENTS frequently believe that mathematical procedures are unique and timeless. They do not realize that over the generations people have grappled with many methods in attempting to find those that were simpler, quicker, less wasteful of resources, or easier to communicate. Further, they don't realize that this search continues. Consequently, students do not see themselves as future inventors of mathematics, an important vision if mathematics is to continue to grow.

One way to address this issue is to have students explore algorithms through individual or small-group projects. Projects are a valuable teaching and learning tool for many reasons (McConnell 1995). Projects—

- are natural opportunities for sharing, communication, and alternative assessment;
- furnish opportunities to multiply the number of topics introduced in a course;
- give students ownership of a task;
- promote the use of students' varied intelligences;
- create avenues through which new mathematical topics can be studied and familiar ideas can be applied, transferred, and deepened.

Finally, projects advance an important general school goal—that of encouraging students to become independent learners.

In addition to these benefits, projects on algorithms have the potential to integrate into school programs the history of mathematics and the contributions of many cultures (Nelson, Joseph, and Williams 1993; Ascher 1991). They help students recognize that mathematics has not always been done in one way and that people around the world and over time have searched for ways to make mathematics easier. With such a perspective, we can help students see that they, too, can continue to contribute to this global legacy of helping mathematics grow.

With an orientation to the history of algorithms, guidance to helpful resources, and some good beginning questions, students can investigate algorithms and share their findings in oral, visual, or written presentations. This paper presents ideas from history for sample topics and springboard questions to help get students started.

EGYPTIAN FRACTIONS

The ancient Egyptians needed fractions because trade was done with goods, not money (Joseph 1992). So barley, land, and other valuables needed to be divided to pay for purchases. The Egyptians did not, however, have the procedures or algorithms with which we are familiar today for operations with fractions. Egyptian scribes calculated fractions using almost entirely unit fractions, which are fractions with a numerator of 1 (Bunt, Jones, and Bedient 1976; Gillings 1972; Joseph 1992). (The one exception was 2/3. It is not clear how this happened.) A fraction was denoted by an open mouth (or a dot) over the symbol for the whole number, as shown in figure 13.1.

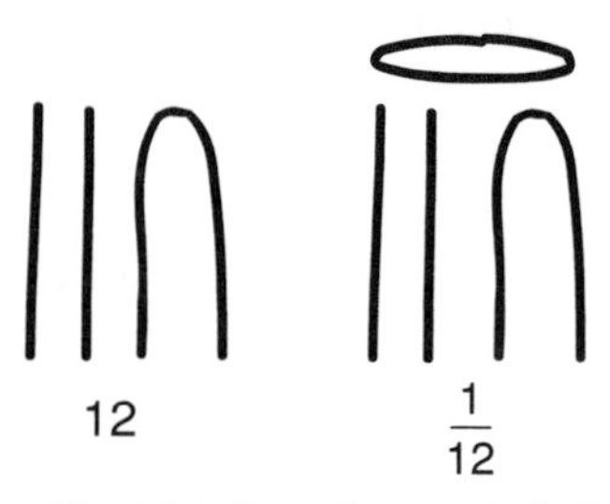

Fig. 13.1. Egyptian symbols for 12 and 1/12

Because the additions, once calculated, could be used again and again, the scribes preserved fraction sums in tables written on scrolls. No addition or equal signs were used. So, for example, on a famous record called the Egyptian Mathematical Leather Roll, symbols for the sentence 1/9 + 1/18 = 1/6 appeared as follows, with bars replacing "mouths":

$$\overline{9}\quad\overline{18}\quad\overline{6}$$

Questions for Investigation

1. Find other examples of unit fractions that give a total of other unit fractions. Record them using the Egyptian notation.
2. Egyptian scribes were very familiar with the two sums:

$$\frac{1}{3}+\frac{1}{6}=\frac{1}{2}$$

$$\frac{1}{4}+\frac{1}{12}=\frac{1}{3}$$

These two sums suggest a possible pattern:

$$\frac{1}{n+1}+\frac{1}{n(n+1)}=\frac{1}{n}$$

Find other examples of this pattern, if possible.

Is this pattern always true? Show your reasoning.

3. Check the following fraction additions as shown on the Egyptian Mathematical Leather Roll (Gillings 1972, p. 39). Find any patterns that you can.

$$\overline{9}\quad\overline{18}\quad\overline{6}$$
$$\overline{12}\quad\overline{24}\quad\overline{8}$$
$$\overline{24}\quad\overline{48}\quad\overline{16}$$
$$\overline{18}\quad\overline{36}\quad\overline{12}$$
$$\overline{21}\quad\overline{42}\quad\overline{14}$$
$$\overline{45}\quad\overline{90}\quad\overline{30}$$
$$\overline{30}\quad\overline{60}\quad\overline{20}$$
$$\overline{15}\quad\overline{30}\quad\overline{10}$$
$$\overline{48}\quad\overline{96}\quad\overline{32}$$
$$\overline{96}\quad\overline{192}\quad\overline{64}$$

4. Gillings, who studied Egyptian scrolls, phrased a pattern he found for the additions in #3 as an Egyptian scribe might: "For adding 2 fractions, if one number is twice the other, divide it by 3" (Gillings 1972, p. 40). (Note that the "numbers" are the unit-fraction denominators and "it" refers to the second number in patterns like those above.) Express Gillings's rule using algebra. If his rule is not true, find an exception. If it is true, prove it and look for an extension.
5. Can all unit fractions be expressed as the sum of two other unit fractions? Justify your answer.
6. What advantages and disadvantages were there to using only unit fractions?
7. What other relationships, properties, or curiosities can you find about unit fractions?
8. Some calculators perform operations with fractions, whereas other calculators use decimals only. What algorithms do you need to know to do fraction operations on a decimal-only calculator?

CIRCLE FORMULAS BEFORE PI

Pi was an elusive concept for a long time. The ratio of a circle's circumference to its diameter refused to lend itself to expression by a whole or rational number. Without irrational numbers, the ancients had only approximate formulas for circles.

Questions for Investigation

1. Why can we think of a formula for the area or circumference of a circle as an algorithm?
2. The Egyptians found the area of a circle with a procedure that was effectively $A = \left(d - \frac{1}{9}d\right)^2$. (See NCTM [1969, p. 126] for more information.)
   a) Use this formula to find the areas of different circles. Compare the results with those you get using $A = \pi r^2$ and a calculator with a multidigit approximation of pi. Estimate the percent error. How well did the Egyptians do without pi?
   b) Rewrite the Egyptian formula so that it can be compared with πr^2. What value were the Egyptians using for π?
   c) Use the Egyptian algorithm to construct a square whose area "equals" that of a circle with a given diameter.
3. The Babylonians thought the circumference of a circle was three times as long as the diameter. For the area of a circle they took, effectively, one-twelfth the area of a square whose side was the length of the circumference (NCTM 1969, pp. 168–69). Write a formula for the Babylonian method. Rewrite it to see how it compares with $A = \pi r^2$.
4. A Hindu writer, Brahmagupta (ca. A.D. 628), gave $\sqrt{10}$ as the "exact value" of pi (NCTM 1969, p. 151). What would have been Brahmagupta's formula for finding the circumference of a circle? The area of a circle? What percent error does Brahmagupta's value produce?
5. Archimedes estimated the area of a circle using inscribed and circumscribed polygons (Dunham 1990). How did his method work?
6. Modern computers can calculate pi to thousands of decimal places. Often, performing such procedures accurately is a test of a computer's accuracy and efficiency. What algorithms are used to produce multidigit representations of pi? How do these algorithms compare in efficiency?

ALGEBRA WITHOUT VARIABLES

People needed to find unknown measures well before the invention of algebraic notation in the fifteenth century. One popular method of that era is known as "false position" (Boyer 1985): A guess was made at an answer, the guess was used to check, and then a revision was made on the basis of the factor by which the guess was wrong. For example, the Ahmes Papyrus contains the following problem (Joseph 1992, p. 78):

A quantity and its quarter added become 15. What is the quantity?

The scribe used this method:

$$\text{Guess 4; then } 4 + \frac{1}{4} \text{ of 4 would be 5.}$$

But the actual result is 15, or 3 times 5. This means the first guess must be multiplied by 3 to be correct. So, $3 \times 4 = 12$, the solution. That is,

$$12 + \frac{1}{4} \text{ of } 12 = 15.$$

Questions for Investigation

1. Write an equation for the original problem using algebra. Solve it using procedures you have learned.
2. Create other simple problems and solve them using proportional reasoning, like the scribe used, and then solve them with variables.
3. Here is another, more complicated problem from the Berlin Papyrus (Joseph 1992, p. 78):

> It is said to thee that the area of a square of 100 square cubits is equal to that of two smaller squares. The side of one is 1/2 + 1/4 of the other. Let me know the sides of the two unknown squares.

The Egyptian solution began with a guess of 1 for the side of one of the smaller squares. Then the other square must have sides of length 1/2 + 1/4. The areas of the squares are 1 and 1/2 + 1/16. (Recall that the Egyptians used only unit fractions!) Adding these together gives 1 + 1/2 + 1/16. Its square root is 1 + 1/4. The square root of 100 is 10. Then 10 divided by the result of 1 + 1/4, which is 8, must be the side of the first smaller square. (The papyrus was tattered and unreadable for the remainder of the solution.)

   a) Verify the solution steps given.
   b) Finish the solution to the Egyptian problem.
   c) Solve the same problem using variables.
   d) Create another, similar problem with squares and solve it by both methods.

4. The "false position" method works using proportions.
   a) Create and solve other problems that can be solved using "false position."
   b) Create other problems that can be solved with algebra but are not solvable by the "false position" method. For what kinds of problems does the "false position" method work?
   c) What are some advantages and disadvantages of the Egyptian method of "false position"?
5. Why is it reasonable to think of procedures for solving algebraic equations as algorithms?

SUMMARY

This paper has offered a rationale for students to investigate algorithms of the past and has detailed some sample starting points. Students can investigate the topics illustrated and other historical algorithms by using tasks like the following:

- Try out the procedure.
- Verify it or find its limitations.
- Represent the method in another form.
- Find out more, if possible, about its inventor or inventors.
- Compare it to current methods.
- Identify advantages or disadvantages of the method.
- Tell your feelings about doing mathematics this way.

It is hoped that explorations like these, orchestrated by insightful teachers and shared by students, will build an appreciation for the ever-changing nature of mathematics and its algorithms.

REFERENCES

Ascher, Marcia. *Ethnomathematics: A Multicultural View of Mathematical Ideas.* Pacific Grove, Calif.: Brooks Cole Publishing Co., 1991.

Boyer, Carl B. *A History of Mathematics.* Princeton, N.J.: Princeton University Press, 1985.

Bunt, Lucas N. H., Philip S. Jones, and Jack D. Bedient. *Historical Roots of Elementary Mathematics.* New York: Dover Publications, 1976.

Dunham, William. *Journey through Genius: The Great Theorems of Mathematics.* New York: Penguin Books, 1990.

Gillings, Richard J. *Mathematics in the Time of the Pharaohs.* New York: Dover Publications, 1972.

Joseph, George Gheverghese. *The Crest of the Peacock: Non-European Roots of Mathematics.* London: Penguin Books, 1992.

McConnell, John W. "Forging Links with Projects in Mathematics." In *Connecting Mathematics across the Curriculum,* 1995 Yearbook of the National Council of Teachers of Mathematics, edited by Peggy A. House, pp. 198–209. Reston, Va.: National Council of Teachers of Mathematics, 1995.

National Council of Teachers of Mathematics. *Historical Topics for the Mathematics Classroom.* Thirty-first Yearbook of the National Council of Teachers of Mathematics. Washington, D.C.: National Council of Teachers of Mathematics, 1969.

Nelson, David, and George Gheverghese Joseph, and Julian Williams. *Multicultural Mathematics: Teaching Mathematics from a Global Perspective.* Oxford: Oxford University Press, 1993.

ADDITIONAL READING

Eves, Howard. *Great Moments in Mathematics before 1650.* Washington, D.C.: Mathematical Association of America, 1983.

Swetz, Frank J., and T. I. Kao. *Was Pythagoras Chinese? An Examination of Right Triangle Theory in Ancient China.* Reston, Va.: National Council of Teachers of Mathematics, 1977.

Zaslavsky, Claudia. *Africa Counts: Number and Pattern in African Culture.* Boston: Prindle, Weber & Schmidt, 1973.

———. *Multicultural Mathematics: Interdisciplinary Cooperative Learning Activities.* Portland, Maine: J. Weston Walsh, Publisher, 1993.

Part 3

14

Alternative Algorithms for Whole-Number Operations

William M. Carroll

Denise Porter

ALTHOUGH the importance of problem solving, conceptual understanding, and the development of number sense has been documented in the various recent reform initiatives (Hiebert et al. 1996; NCTM 1991), it is time to evaluate the level of prominence given to computation and the standard algorithms. Classrooms are envisioned in which students are encouraged to develop and use a variety of approaches rather than a prescribed algorithm for performing computations. When faced with a computation, a student might select a manipulative to model the situation, develop an invented procedure, draw a picture, perform a mental calculation, choose an appropriate written algorithm, or use a calculator. For a relatively simple calculation, like 35 – 17, a mental procedure, such as adding up, might be most appropriate. For more complex calculations, such as 36 × 58, students might be more likely to use a calculator, apply some written algorithm, or use an estimation technique if an exact answer is not necessary. Instead of focusing solely on the answer, students are encouraged both to reason about which approach might best fit the problem at hand and to consider the reasonableness of their answer.

Providing opportunities for students in the primary grades to develop, use, and discuss invented algorithms helps to enhance number and operation sense (Kamii, Lewis, and Livingston 1993; Sowder 1992). In our observations of classrooms and interviews with students where this approach is used, we have found students to be more flexible in their thinking and in choosing a procedure that best fits the problem. Often, they are aware of several ways to solve the problem but choose the most appropriate. When encouraged to share their strategies, students develop better reasoning and communication skills. They are also exposed to a wider range of solution methods (Carroll and Porter 1997).

These results suggest that the teaching of specific algorithms should perhaps be diminished or delayed in the early primary school curriculum (Burns 1994). However, as students move beyond the primary grades, written algorithms are still useful, for several reasons. First, some calculations are not handled as well by a mental method. As numbers become greater, demand on memory grows and often requires written record keeping. Although calculators and estimation can often be used, written algorithms are often the most reasonable method to use for "middle sized" numbers. Second, some students are not successful at "inventing" addition and subtraction algorithms, and some alternative, like the traditional written method, may be beneficial to them. Methods that build on their thinking and early experiences can help them use written algorithms with meaning. Third, tests and other societal expectations still make it necessary for most students to know some written method for each operation. Not only do many standardized tests still include a computation section, but parents and other adults still see competence in written computation as a benchmark of success in mathematics. Finally, students benefit from having a range of options to choose from, including paper-and-pencil methods. This is supported by our observations of fourth-, fifth-, and sixth-grade classrooms where students are introduced to several methods for the subtraction and multiplication of whole numbers.

Although it is advantageous for all students to know at least one written procedure for each of the operations, the standard algorithms taught in school are often not the most appropriate or understandable. Although they are efficient, the meaning of these standard algorithms (e.g., why the "ones" are being carried) is often unclear to students who learn them without understanding. Consequently, "buggy" procedures may develop. This paper suggests some alternative algorithms that have been used successfully in school. Generally, these algorithms reflect students' natural solution procedures, such as left-to-right addition or repeated estimations for division. Most are analogous to the mental or manipulative-based procedures that students develop, but they allow them a paper-and-pencil method for working on larger numbers. Unlike the traditional written algorithms, these algorithms explicitly note place value. They also avoid such confusing procedures as "borrowing" or "carrying" and generally separate the operations within an algorithm, thus allowing an option for students who have learned, but are confused by, the standard algorithms.

Alternative written algorithms for each of the four operations are discussed below. These algorithms are introduced in the University of Chicago School Mathematics Project elementary school curriculum and similar reform curricula after the students have had time to explore and devise their own methods, generally in third grade or later. Although these are not the only options, these are the ones with which we have found students to be successful. These particular algorithms are also well liked by teachers because students often show more understanding and are often more likely to be successful using them. Of course, these algorithms should be part of an instructional scheme that includes estimation, mental arithmetic, calculators, student-invented methods, and discussions of the different strategies used.

AN ADDITION ALGORITHM

Most students develop some effective method for solving addition problems. For example, in solving 27 + 56, many students will mentally add 20 + 50 = 70, then 6 + 7 = 13, and finally 70 + 13 = 83. Other students may reformulate the problem as 25 + 50 + 2 + 6 = 75 + 8. However, as numbers become greater or contain decimals, mental computation becomes difficult. A method that many teachers and students have found useful is the *partial sums* method. As in the first mental solution above, numbers are first added by their place value. No regrouping or "carrying" is done until the end, when these partial sums are added together. For example, in figure 14.1, the student first added the thousands (2000 + 1000) and then continued from left to right, recording each partial sum. Finally, these sums are combined. Base-ten blocks can be used to model the additions for younger students or those who are having difficulty linking the operation to the written symbols.

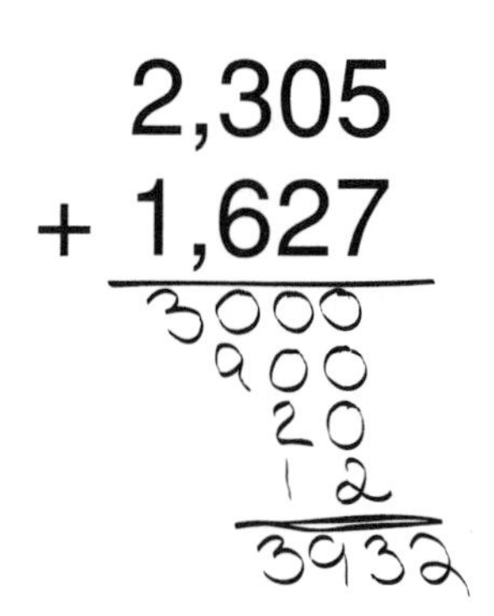

Fig. 14.1. Partial sums for addition

This method has the advantage of emphasizing place value. That is, students are not treating the columns as stacks of units, adding 5 and 7 and carrying the "one." Instead, they are adding 5 ones and 7 ones and recording it as 12 (fig. 14.1). This method nicely leads into the partial differences and partial products that are illustrated later in this paper.

SUBTRACTION ALGORITHMS

Because more students have difficulty developing an accurate and understandable subtraction algorithm, two methods that we have seen successfully used are suggested here.

Adding up

The *adding up* method is an algorithm that many students "invent" on their own in first or second grade, usually as a mental procedure for working with smaller numbers. This procedure focuses on the comparison meaning of subtraction or subtraction as the inverse of addition. For example, in the problem 41 – 25, the students ask, "How far apart are 41 and 25?" or "What do I need to add to 25 to get 41?" This can be solved by adding up in steps: 25 + 10 is 35; 35 + 5 is 40; and 1 more is 41 (see fig. 14.2a). The difference between the two numbers is the sum of the numbers added: 5 + 10 + 1 = 16. An alternative method for recording the "steps up" is shown in figure 14.2b, where the student was solving the problem 124 – 79.

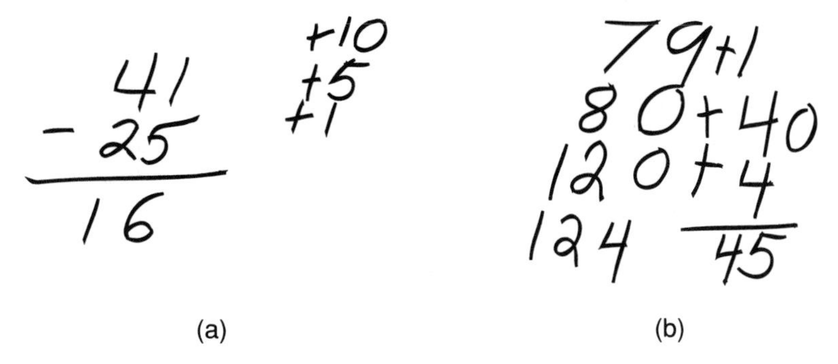

Fig. 14.2. Two methods for recording "adding up" for subtraction

People often develop an adding-up procedure as a replacement for subtraction. For example, counting up is the method many of us use for making change during a purchase. Thus some students who are having trouble developing a subtraction algorithm or using the standard method successfully may benefit from being shown the written method for adding up, especially as numbers become too large to handle mentally.

This algorithm has several strengths. First, it builds on the children's number sense ("What should I add to 79 to get to 100?"). Second, it supports the meaning of subtraction as the comparison of two numbers. For this reason, this algorithm can be used later to introduce the subtraction of signed numbers in a nonalgorithmic manner. For example, in solving 25 – (–13), a student can ask, "What should I add to –13 to get to 0?" and "What do I add to 0 to get to 25?" With the support of a number line, this is simple.

Partial differences

This algorithm is similar to the partial sums algorithm for addition. In this method (see fig. 14.3), the difference between the two numbers in each column is found, emphasizing the place value, that is, "5 hundreds minus 2 hundreds" rather than "5 minus 2." Unlike the standard algorithm, subtractions usually proceed from left to right, and if the subtrahend is larger, the difference is recorded as a deficit or a negative number. For example, in figure 14.3, the student recorded the difference 40 – 70 as –30. Thus, no "borrowing" or regrouping is necessary. Finally, the partial differences are combined, in steps if necessary: (300 – 30) – 8.

Although the use of negative numbers may seem difficult for elementary school students, many use them in this context with little difficulty, and some primary school students develop this method on their own. It is not important whether students actually think of these partial sums as negative and positive numbers. They may simply consider the negatives as "being in the hole" or having a deficit of that quantity.

MULTIPLICATION ALGORITHMS

Even before learning the multiplication facts, many students develop a repeated-addition method for multiplication (e.g., $4 \times 7 = 7 + 7 + 7 + 7$). Efficient as this is for smaller numbers, repeated addition becomes cumbersome for two-digit and larger numbers. Unlike addition and subtraction, few students invent an efficient algorithm for multidigit multiplication. Two alternatives to the standard multiplication algorithm are discussed below.

Partial products

This method is similar to the *partial sums* and *partial differences* algorithms discussed above. Often the multiplications are done left to right, although in figure 14.4, the student began with 4×7 (28), followed by 20×7 (140), 30×4 (120), and 30×20 (600). Like the addition and subtraction algorithms discussed earlier, this method emphasizes the place value of the multipliers and the partial products. It also eliminates "carrying" during the course of multiplication, leaving all regrouping until the end.

To be successful with this algorithm, students must not only know the basic facts but be able to extend them to 20×7 or 20×30. Since this is an important skill in estimation (e.g., 51×72 is about 50×70), most students should have already developed some facility with these extended facts. This algorithm can also be extended to the multiplication of mixed numbers.

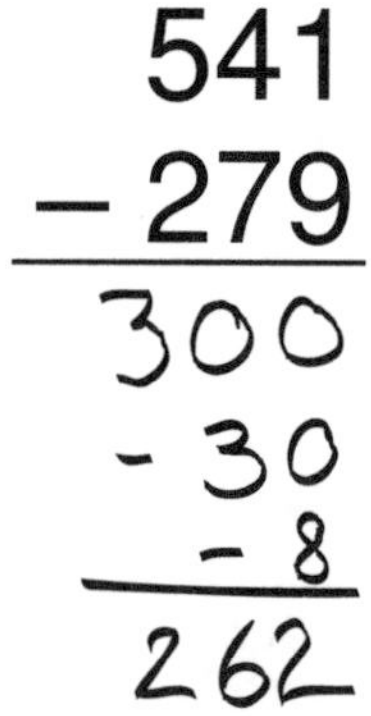

541
− 279
300
- 30
- 8
262

Fig. 14.3. Partial differences for subtraction

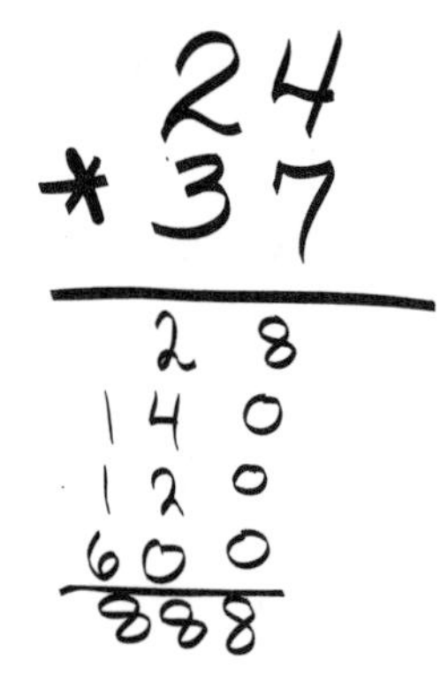

Fig. 14.4. Partial-products method for multiplication

Lattice multiplication

Historically, the first records of this algorithm date back to tenth-century India (Nelson, Joseph, and Williams 1993). Along with the Hindu-Arabic numeration system, this algorithm was imported to Europe and was popular during the fourteenth and fifteenth centuries. Napier's rods (or "bones"), a multiplication procedure developed by John Napier in the early 1600s, were modeled on lattice multiplication.

Although lattice multiplication ignores place value, it has the advantage of delaying addition (and "carrying") until all multiplications are complete. The procedure is illustrated and described in figure 14.5. Although the reasons are not obvious to us, this method has proved to be very popular with students. Low-achieving students especially seem to like this algorithm, perhaps because of the structure provided by the lattice. In our observations of classrooms where students are introduced to alternative algorithms, we have found students using all three multiplication methods: the standard written algorithm, the partial-products method, and the lattice method. We have also found that many students know at least two of these methods. Although it may not be necessary that they know more than one approach, many students seem to approach problems with more confidence knowing that if one approach does not work for them, they have another procedure that they can use.

In lattice multiplication, the multipliers, in this example 24 and 37, are written along the top and right sides. Each partial product is recorded in the corresponding cells. For example, $4 \times 3 = 12$ is recorded in the upper right cell, and $2 \times 3 = 06$ is recorded in the upper left cell. When all multiplications are complete, numbers are summed along the diagonals.

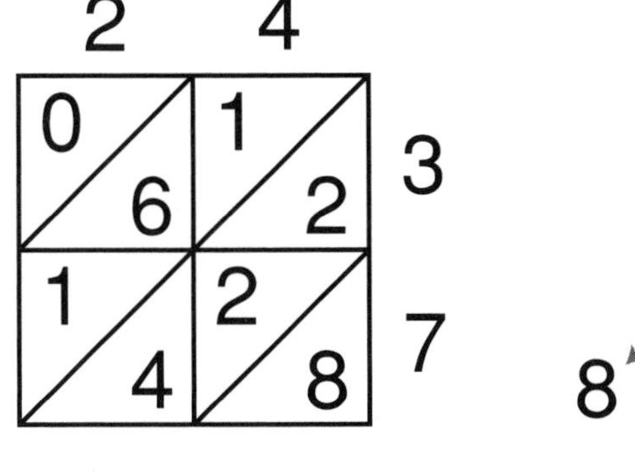

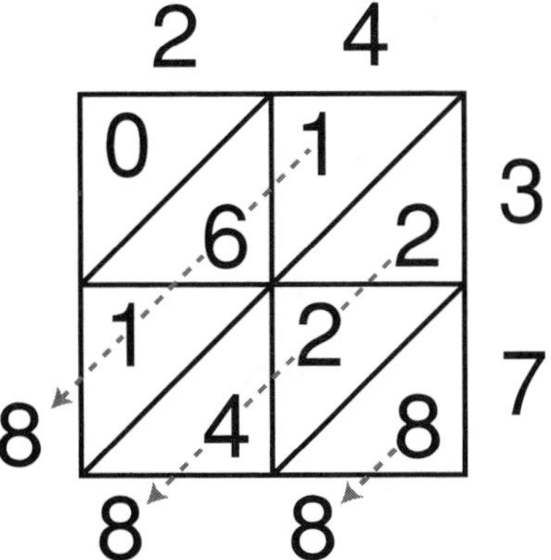

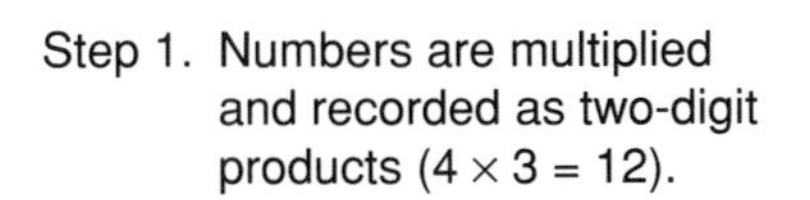

Step 1. Numbers are multiplied and recorded as two-digit products ($4 \times 3 = 12$).

Step 2. These partial products are summed along the diagonals.

Fig. 14.5. The lattice multiplication algorithm for 24×37

AN ALTERNATIVE LONG-DIVISION ALGORITHM

The traditional long-division algorithm is difficult for many students. Many never master it in elementary school, and fewer develop meaning for the procedure or the answer (Silver, Shapiro, and Deutsch 1993). Although there are several reasons for this, two stand out. First, the standard algorithm requires students to get an exact answer in the quotient at each step. But in real-life division situations, children often move in steps toward the solution. For example, in solving a problem like "There are 50 marbles to be divided among 8 children," a student may make a first guess of 5 marbles each, or 40 marbles. At this point, it should be clear that of the remaining marbles, each child can receive 1 more (which makes a total of 48), with 2 marbles left over. Second, the standard algorithm requires students to ignore place value. In dividing 847 by 9, the student asks, "How many times does 9 go into 84?" rather than "How many 9s can I get out of 847?" For these reasons, the standard algorithm works against students' number sense.

The repeated-subtraction algorithm illustrated in figures 14.6a and 14.6b instead builds on number sense and problem-solving skills. Students estimate about how many times a divisor can be subtracted from the dividend, and each estimate brings the students closer to the solution. The solutions of two fifth graders to the problem of dividing 847 by 9 are shown in figure 14.6. Rather than recording an exact quotient at the top, students first make repeated approximations, reducing the next dividend each time until it is less than the divisor. The algorithm also allows students to solve the problem at their own level of achievement. Although the second student (fig. 14.6b) took more steps than the first, he or she still arrived at the correct answer.

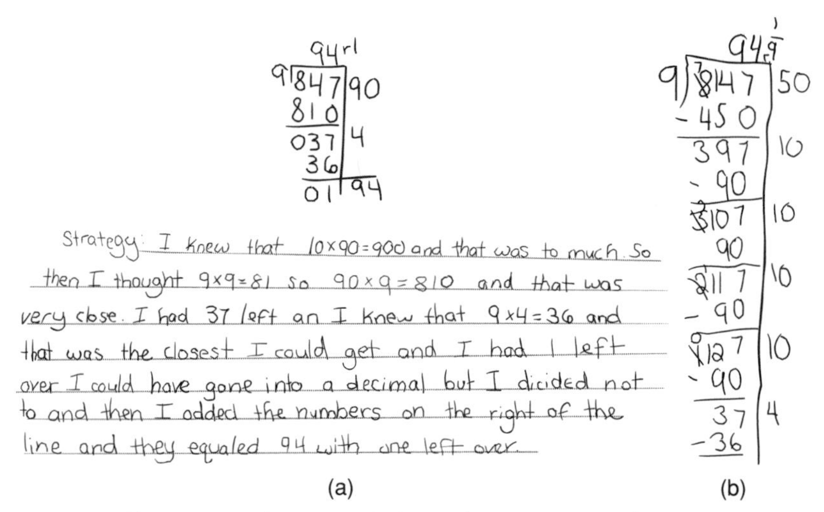

Fig. 14.6. An alternative division algorithm: two solutions

Several teachers have noted that this algorithm is much easier than the standard algorithm for students to use and understand. As one teacher reported:

> Students took a while to get used to this method, but once they caught on, they were really excited. For many, this was the first time that they could ever be successful with long division.

As illustrated in figure 14.6a, many teachers find it useful to have students write about the algorithms that they use. These writings may be incorporated into a journal for inclusion in a student's portfolio or as a letter to a parent or next year's teacher. This encourages students to think through the steps of their algorithms and clarify and refine their ideas about the procedures they used. It also furnishes an opportunity to incorporate writing into the mathematics class.

Students' writing about their algorithms can be used not only to assess their communication skills but also to evaluate their understanding and use of the algorithm. Written descriptions of the algorithm can help to clarify for the teacher whether the student merely made a careless slip or has a deeper misunderstanding of the procedure—something that may not be clear from the use of the algorithm alone. Some teachers have found it helpful to incorporate this writing into students' letters describing their algorithms to their parents. Because most parents know only the traditional algorithms, this clarifies the mathematics their child is learning and gives them a better opportunity to help at home. Finally, this same method can be used to introduce next year's teacher to the algorithms and strategies that students learned and used this year.

CONCLUSION

The algorithms presented in this paper provide nice alternatives to the traditional procedures. One reason these alternatives are often successful is that they are more similar to students' invented methods, building on their own thinking about whole-number operations. Important mathematical ideas like place value and estimation can be enhanced as students use these algorithms. Although the examples presented here all involve whole numbers, some of these algorithms can be extended to decimal numbers, and similar alternative algorithms can be found for fractions.

These alternative algorithms might be useful in class instruction in several ways. For students who have been through one of the reform curricula that encourage invented algorithms and the use of calculators, these algorithms provide a pencil-and-paper method for working with midsized numbers that are not so easily computed mentally but for which a calculator is not necessary. For students who have learned the traditional algorithms in earlier grades, these algorithms can be presented as alternative ways for solving problems—for example, counting up is a much more reasonable approach for solving 7 000 – 6 925. Presenting alternative algorithms also illustrates that the traditional algorithms are only one of several possible methods for solving problems. Discussions of these algorithms may also lead to fruitful discussions of students' invented methods as well as algorithms used in other countries. Finally, for students who are struggling with traditional algorithms, some of these can be presented as alternative methods. For example, because the lattice method delays addition until the end, students who have difficulty with "carrying" may find this method an easier one. In any event, these algorithms should be used in conjunction with mathematics instruction that stresses estimation, problem solving, and the understanding of the operations involved in computation.

REFERENCES

Burns, Marilyn. "Arithmetic: The Last Holdout." *Phi Delta Kappan* 75 (1994): 471–76.

Carroll, William M., and Denise Porter. "Invented Strategies Can Develop Meaningful Mathematical Procedures." *Teaching Children Mathematics* 3 (March 1997): 370–74.

Hiebert, James, Thomas P. Carpenter, Elizabeth Fennema, Karen Fuson, Piet Human, Hanlie Murray, Alwyn Olivier, and Diana Wearne. "Problem Solving as a Basis for Reform in Curriculum and Instruction: The Case of Mathematics." *Educational Researcher* 25 (May 1996): 12–21.

Kamii, Constance, Barbara A. Lewis, and Sally Jones Livingston. "Primary Arithmetic: Children Inventing Their Own Procedures." *Arithmetic Teacher* 41 (December 1993): 200–203.

National Council of Teachers of Mathematics. *Professional Standards for Teaching Mathematics.* Reston, Va.: National Council of Teachers of Mathematics 1991.

Nelson, David., George G. Joseph, and Julian Williams. *Multicultural Mathematics.* Oxford: Oxford University Press, 1993.

Silver, Edward A., Lora J. Shapiro, and Adam Deutsch. "Sense Making and the Solution of Division Problems Involving Remainders: An Examination of Middle School Students' Solution Processes and Their Interpretations of Solutions." *Journal for Research in Mathematics Education* 24 (March 1993): 117–35.

Sowder, Judith. "Estimation and Number Sense." In *Handbook of Research on Mathematics Teaching and Learning,* edited by Douglas A. Grouws, pp. 371–89. New York: Macmillan Publishing Co., 1992.

15

My Family Taught Me This Way

Pilar Ron

I REMEMBER the first time I saw subtraction done by using the algorithm taught in American schools. I was reading a paper on the typical mistakes children make when learning multidigit subtraction, and I felt like one of those children, trying to make sense of the numbers in front of me. I wondered what those crossed-out numbers were and what the author of the paper meant by *ignored borrows*. I had to read the paper three times before I realized that my way of doing subtraction was very different from that of the author of the paper and of the children making those typical mistakes. There I was, a reasonably intelligent adult woman, and yet I could make sense of the American subtraction algorithm only with the aid of a paper. Since then I have met many other people who have shown an equal lack of understanding when faced with multidigit subtraction as solved by American-schooled children. Those people and I did not necessarily share a language or a culture, but we did share a common trait: we had all been schooled in Europe or in South America.

I have spent the last three years working on a grades K–3 mathematics research project in bilingual inner-city schools in which 88 percent of the children have a Latino background (see Fuson, Feingold, and Cuevas [1996] for details). Our approach is to allow children to use different methods with the support of base-ten materials for understanding. However, we have worked with teachers who teach the standard U.S. algorithm and have seen the confusions that result when parents that were schooled in other countries teach their children their "add tens to both" method (a method found in most European and Latin countries, among others). Because parents will always try to help their children with mathematics as they know it, it is helpful for teachers to understand the European-Latino algorithm (E-L algorithm), the errors that children make with it, and the confusions that arise from mixing the U.S. and the E-L algorithms.

THE ALGORITHMS

A common U.S. algorithm for subtraction is demonstrated in figure 15.1: Borrow one ten to go with the ones, that is, regroup the top number.

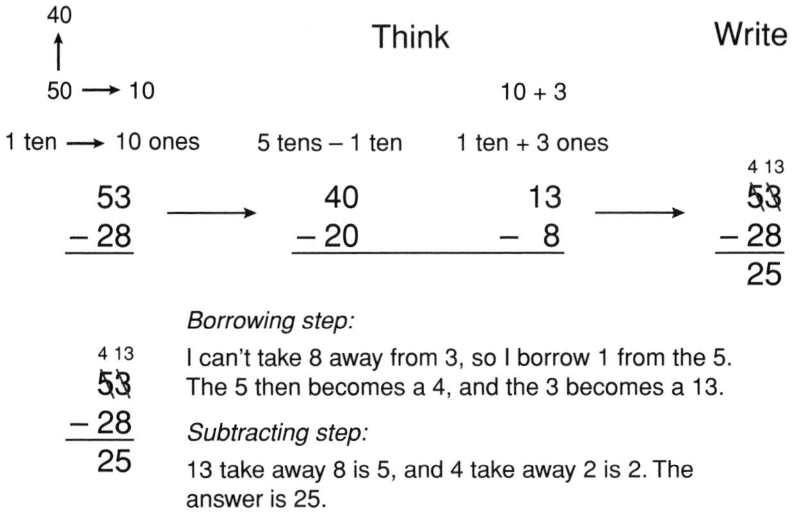

Borrowing step:

I can't take 8 away from 3, so I borrow 1 from the 5. The 5 then becomes a 4, and the 3 becomes a 13.

Subtracting step:

13 take away 8 is 5, and 4 take away 2 is 2. The answer is 25.

Fig. 15.1. The U.S. subtraction algorithm

The E-L algorithm for subtraction is shown in figure 15.2: Add a ten to both numbers. This algorithm, which is also known as the *equal additions method,* is found in many countries, including most European and South American countries. The E-L algorithm relies on the fact that the result of subtracting 28 from 53 is the same as that of subtracting 38 from 63. To benefit from this mathematical fact, both numbers are changed equally by adding a ten to each number. First add 1 ten to the ones in the top number (the minuend), so 3 becomes 13. Then add the other ten to the tens in the bottom number (the subtrahend), so 2 tens becomes 3 tens. The actual subtracting is usually done as counting up (i.e., rather than "13 take away 8 is 5," the child says or thinks "from 8 to 13 is 5"). The adding of a ten to the subtrahend is expressed by the phrase "I carry 1," which is the same wording used in the addition algorithm when a 1 (actually a ten) is carried over. For users of the E-L algorithm, the similarity of expressions opens the door to the potential confusion of addition and subtraction. Typically one thinks of putting a 1 in front of the 3 to make 13 and then carrying 1 into the tens column in the subtrahend instead of thinking of adding a ten. As with the U.S. algorithm, most parents do not understand why the E-L algorithm works or that it really adds ten to *each* multidigit number.

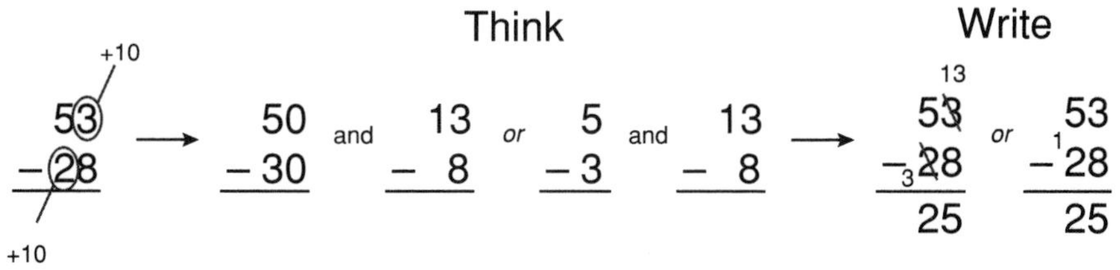

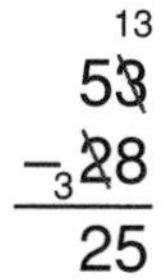

		Think				Write	
$\begin{array}{r} 5\textcircled{3} \\ -\textcircled{2}8 \\ \hline \end{array}$ (+10 / +10)	→	$\begin{array}{r} 50 \\ -30 \\ \hline \end{array}$ and $\begin{array}{r} 13 \\ -8 \\ \hline \end{array}$	or	$\begin{array}{r} 5 \\ -3 \\ \hline \end{array}$ and $\begin{array}{r} 13 \\ -8 \\ \hline \end{array}$	→	$\begin{array}{r} 5\overset{13}{\not 3} \\ -_3\not 28 \\ \hline 25 \end{array}$	or $\begin{array}{r} 53 \\ -{}^{1}28 \\ \hline 25 \end{array}$

$$\begin{array}{r} 5\overset{13}{\not 3} \\ -_3\not 28 \\ \hline 25 \end{array}$$

Add ten to the ones column in the minuend:
I can't take 8 away from 3, so I make the 3 into a 13.

Subtract the ones:
From 8 to 13 is 5.

Add ten to the tens column in the subtrahend:
I carry 1; 2 and 1 (the 1 I carry) is 3.

Subtract the tens:
From 3 to 5 is 2. The answer is 25.

Fig. 15.2. The European-Latino subtraction algorithm

TYPICAL ERRORS AND CONFLICTS

Both algorithms may be used by students in a classroom even if a teacher does not teach them because they may be taught in the homes. Because some parents emphasize working mentally, teachers may not see marks on paper that can be used to diagnose errors that students may be making. Asking students to explain their methods can help the teacher uncover the reasons for the errors that do arise.

A child who has been exposed to both the U.S and the E-L algorithms may make mistakes that stem from taking characteristic elements from both. The phrase "my family taught me this way" is often heard in the classroom as an explanation for a procedure that students realize was not taught in class. Teachers need to understand the reason for the potential mistakes so that they can provide help wherever it is needed.

The two most common mistakes that result from mixing the two subtraction algorithms are shown in figure 15.3. Some children begin with the U.S. borrowing algorithm, then switch algorithms during the move to the tens column. They "carry 1" but in doing so shift to some form of addition. In version 1 they shift completely to addition and end up subtracting in the ones column and adding in the tens column. In version 2 they do not shift to adding the tens column but add the ten to the minuend instead of to the subtrahend before subtracting in the tens column. (These examples are not the work of any one child but represent recurring mistakes that we have found among lower-achieving children of Latino background.)

Mixed U.S.-E/L Version 1

(the 1, 4, and 13 may or may not be written)

```
   1
  4 13
  5 3
- 2 8
-----
  7 5
```

Borrow to subtract ones (U.S.):

I can't take 8 away from 3, so I borrow 1 from the 5. The 5 then becomes a 4 and the 3 becomes a 13.

Subtract ones (U.S.):

13 take away 8 is 5.

Carry the 1 (E/L):

I carry 1.

E/L error: Shift to addition so the 1 is added to the tens column in the minuend instead of in the subtrahead, and the entire tens column is then added:

The 1 I carry and 4 is 5 and 2 is 7. The answer is 75.

Mixed U.S.-E/L Version 2

(the 1, 4, 5, and 13 may or may not be written)

```
 1 5
   4 13
   5 3
-  2 8
------
   3 5
```

Borrow to subtract ones (U.S.):

I can't take 8 away from 3, so I borrow 1 from the 5. The 5 then becomes a 4 and the 3 becomes a 13.

Subtract ones (U.S.):

13 take away 8 is 5.

Carry the 1 (E/L):

I carry 1.

E/L error: Shift to addition so the 1 is added to the tens column in the minuend instead of in the subtrahead, and then the subtraction is performed in the tens column:

The 1 I carry and 4 is 5; 5 take away 2 is 3. The answer is 35.

Fig. 15.3. Common errors resulting from mixing the U.S. and the European-Latino subtraction algorithms

The result of the error in version 2 is the same as that obtained if the child had done the borrowing in the ones column but failed to follow through in the tens column. If the student does the procedure mentally, the teacher has no way of knowing if the child forgot to do the borrowing, did the borrowing and then ignored it, or did something else (as in this example). A correction like "You forgot to borrow" would not be of use because the child did not forget. A teacher needs to ask the child how she or he did the subtraction to discover the source of the error.

Although all the foregoing errors are certainly possible—and to some degree common—among lower-achieving children who use the E-L algorithm in American schools, the typical progression is for them to abandon the algorithm they were taught at home and become proficient in the one taught at school if a standard algorithm is taught. However, some children remain confused for a long time. A better understanding by both teachers and parents of the source of the confusion could and should lead to an early correction of the procedure for solving multidigit subtraction.

Conclusions

In a classroom where mathematics is taught with an emphasis on understanding, the U.S. algorithm shows a clear advantage over the E-L algorithm because teachers feel it can be explained meaningfully. A child can be helped to understand that borrowing is actually a regrouping of 53 (5 tens and 3 ones) into 40 + 13 (4 tens and 13 ones). In the current worldwide trend of teaching mathematics for understanding, the U.S. algorithm may be taught in European and Latin American schools in the near future. (In some teachers colleges in Spain, the U.S. algorithm is already being taught to preservice teachers as a preferred alternative to the traditional E-L method.) When and if that happens, teachers from European and Latin American countries will benefit from understanding the two different algorithms and the errors associated with the unexpected mixing of the algorithms that result when children get conflicting messages at school and home.

In a country like the United States where the student population can be so diverse, it is important that teachers find out what knowledge students from all backgrounds bring from their cultures and their homes and *accept it* rather than try to impose their own. Understanding how other algorithms work can give teachers the necessary resources to uncover the reason for certain mistakes that at first glance are unexplainable. Allowing alternative algorithms into the classroom would give voice and value to the knowledge coming from students' homes and has the potential to introduce other mathematical ideas that can be pursued.

Reference

Fuson, Karen, Cathy Feingold, and Saúl Cuevas. "Children's Math Worlds: A Teaching/Learning Project to Support Urban Latino Children's Construction of Mathematical Understanding." Paper presented at the 1996 annual meeting of the American Education Research Association, New York, 8–12 April 1996.

16

Calculators in Primary Mathematics
Exploring Number before Teaching Algorithms

Susie Groves

Kaye Stacey

FORTY years ago, when both authors were in elementary school, the reason for learning standard, written, computational algorithms was obvious: paper-and-pencil calculations were needed for use in everyday life by most members of the community. Being able to perform these algorithms quickly, neatly, and with a high degree of accuracy was a valuable skill for employment. Now that hand-held calculators and calculating devices built into machines such as cash registers have been readily available for two decades, the everyday need for people to become proficient with paper-and-pencil algorithms has virtually disappeared.

The 1990 National Research Council report *Reshaping School Mathematics* recommended a "zero based" approach to curriculum development. In such a curriculum, every topic needs to be justified on its own merits with no area being immune from scrutiny (p. 38). We believe that the amount of time given to paper-and-pencil algorithms should now be the subject of intense debate and experimentation.

In contrast, we believe that there is no need to debate the proposition that all members of society need to have highly developed number sense, including an understanding of number notation, the confidence and the ability to

The Calculators in Primary Mathematics project was funded by the Australian Research Council, Deakin University, and the University of Melbourne. The project team consisted of Susie Groves, Jill Cheeseman, Terry Beeby, Graham Ferres, Ron Welsh, Kaye Stacey, and Paul Carlin.

compute mentally, the ability to select appropriate operations and appropriate computational tools to carry out calculations, and the ability to interpret answers and check their reasonableness. Today's mathematics curriculum needs to give prominence to activities that promote the development of number sense, regardless of whether calculations are being carried out mentally, with paper and pencil, with a calculator, or with computer software.

In this paper we give a range of examples that illustrate how elementary school teachers can begin to develop children's number sense through the use of calculators before standard algorithms are taught. The incidents we report arose in a research project in which elementary school children were given free access to calculators from their first days at school. Examples of the ways in which the children were able to explore number concepts and computational procedures are given, together with a brief summary of the results of extensive written tests and interviews comparing children with and without long-term exposure to calculators.

THE CALCULATORS IN PRIMARY MATHEMATICS PROJECT

The activities described below were developed by teachers of kindergarten to grade 2 in the Calculators in Primary Mathematics project, a long-term investigation into the effects of the use of calculators on the learning and teaching of primary school mathematics. The project, which commenced with kindergarten and first grade in 1990, involved approximately one thousand children and eighty teachers in six schools in Melbourne, Australia. The project followed the children through the schools to fourth grade in 1993, with new children joining the project every year on their entry into school.

The teachers were not given activities or a program to follow but were regarded as part of the research team discovering how calculators could best be used. Feedback and support were provided through visits by members of the project team and by sharing and discussion at regular meetings of teachers and through the project newsletters, which provide a comprehensive record of classroom activities (see Groves et al. [1994] for a published version). As in the Calculator-Aware Number (CAN) project in the United Kingdom (Shuard 1992), teachers devised, shared, and adapted a wide range of excellent activities, which were relevant to their curriculum, flexible, and frequently open-ended and thus catered to a wide range of individual differences in their classrooms.

Counting Activities

One of the most powerful ways in which calculators were used with young children was to support counting. The built-in constant function on the calculator allowed children to begin with any number and skip count by any number they chose. For example, when the student presses [5] [+] [=] [=] [=] [=], the calculator displays in succession 5, 10, 15, 20—a sequence that children can continue. Similarly, when the student presses [3] [+] [5] [=] [=] [=] [=], the display shows 8, 13, 18, 23. We believe that the counting activities developed the children's understanding of number and gave them a feel for the size of large numbers, which is essential for any sensible use of algorithms.

One kindergarten teacher initiated the activity "number rolls," which became popular with many project teachers. Long vertical strips of paper were used to record the results of counting or skip counting using the calculator's constant function. Many children began by counting by ones and continued to do so over the weeks, enjoying the challenge of reaching larger and larger numbers. Others, however, quickly moved on to skip counting by numbers such as 5, 10, or 100. One child, Alex, stated that he was going to count by tens to 1050. How did he know, at age five, that 1050 would be in the sequence? Another child looked at her long number roll and observed that counting by nines usually leads to the units digit's decreasing by 1 each time and the tens digit's increasing by 1. After several weeks, the teacher noticed that many more children were beginning to make conscious predictions about the next number in their sequence—even when they could not read the numbers aloud.

Many kindergarten children used their calculators to count into the thousands and tens of thousands, whereas others counted backward. One day, Daniel wanted to show the project visitor how he could count by fifties. Very quickly he reached 64 250, which he read aloud without hesitation. Ben, who had counted up to 17 900 by hundreds on his number roll, was asked what number would be reached after he pressed equals two more times. He wrote 18 100, although he read it as "eighteen hundred and one."

Simon, also in kindergarten, decided to count by ones starting at 1 000 000. Another child challenged him to reach "one million one hundred." At first, Simon said that there was no such number, but as he got to 1 000 079 he began to think that perhaps there was. When he finally reached 1 000 102 he was thrilled to see that he had "gone right past it."

Alex using a number roll to record counting to 1050 by tens

The students' use of calculators enabled counting to assume a much more prominent place in mathematics lessons. All the teachers continued to use manipulatives as the basis for their mathematics teaching, but counting was no longer tied to the small range of numbers that can be displayed with concrete materials. Nor was counting slow and boring when the numbers became too long to say aloud easily.

With calculators, large numbers, decimals, and negatives all became available through counting. Counting backward led some students to encounter negative numbers. In one kindergarten class in which the children had been discussing and drawing "what lives underground," Alistair said, "Minus means you are going underground." When asked what would be the first number above the ground, he said zero. In another kindergarten class, Kylie confidently used her calculator to count backward and recorded her findings (see fig. 16.1).

Other first- and second-grade students used their calculators to count by decimal numbers. One girl counted aloud as she used the constant function, saying, "Point one, point two, ..., point nine, point ten." She noticed that the display on her calculator showed something different and worked out why.

All these students (but certainly not all children in the project) exhibited a knowledge of number advanced far beyond what is normally expected at their grade levels. For many teachers one of the frightening aspects of calculator use is that students may encounter very large numbers, negative numbers, and decimals "before they are ready." The teachers who were comfortable with calculators in their classroom took the opposite view: they thought their previous curricular constraints had imposed artificial boundaries for the children. The students developed building blocks to use later for more-complex computations. Skip counting with calculators revealed patterns for adding single-digit numbers to other larger numbers, as well as the structure of our place-value numeration system.

Place-Value Activities

Place value underlies all computational techniques that do not rely solely on counting. The teachers were impressed by the contribution calculators made to developing ideas about place value. In many instances, the activities used the calculator simply as a recording device and thus exploited the spontaneous way in which children used it as a convenient "scratch pad" to record information such as telephone numbers.

Fig. 16.1. Kylie's "underground" numbers

For example, one class played “happy families.” Children entered two-digit numbers on their calculators and grouped themselves according to their “families” of tens. They ordered themselves in the families and eventually formed a line around the room from the smallest to the largest number.

Another teacher devised an ordering activity, “number lineup.” A small group of children entered numbers of their choice into their calculators and then ordered themselves at the front of the classroom according to their numbers. More and more children were added to the “lineup,” with new children finding their correct positions. Some students chose negative or very large numbers, which exhibited sophisticated knowledge of the number system. This activity does not use the calculating features of the calculator—it can be done with numbers written on paper—but the calculator makes it easy to play. One day, the same teacher asked the children to enter “the largest number you can read correctly.” Peter, from grade 2, was immediately heard to say, “Ninety-nine billion is two nines, three zeros, another three zeros, and another three zeros.” He wrote 99 000 000 000 on paper because he “couldn’t fit it on the calculator.”

The teacher who devised “number lineup” has used it, and other similar activities, over and over again. She commented that she now prefers activities in which the children can “take themselves where they want to go” and that she is never sure exactly what mathematics is going to happen. She has identified this new preference as a major change in her own teaching of mathematics.

Peter showing the biggest number he could read—99 000 000 000

Having the additional technological resource of the calculator has helped her use many more open-ended activities than in the past.

Computation Activities

Calculators were frequently used as computational tools but were rarely used just to find answers for their own sake. Rather, they were used to solve real-life problems or to investigate numbers. The examples below illustrate how the children used their own invented strategies as well as calculators for computation. But they also show that the children are slow to realize the power of using the four operations, even when they can be accessed at the press of a button. Learning to appreciate the use of the appropriate operations is a long-term endeavor, which has sometimes been overshadowed by the need to practice written algorithms.

In one kindergarten classroom, the children sorted the teddy bears they had brought to school according to color and counted how many there were all together. Some children counted aloud by ones or twos; others used the constant function. A few used addition on the calculator to find the total of the groups—for example, 7 + 4 + 3 + 4 = 18. When recording what they had done, some children drew bears and others wrote number sentences. Byron (see fig. 16.2) immediately wrote the number sentence above correctly and then found other numbers that summed to 18—for example, 9 + 1 + 8, and, even more remarkable, 6 + 6 + 6. Although the calculator was adding for him, he needed to predict which numbers would give a sum of 18 and used the calculator only to test his predictions. Having observed the other number sentences, Byron had concluded that it was likely that 6 + 6 + 6 equals 18 from the other patterns he had seen. This activity encouraged exploration and provided opportunities for the development of number sense.

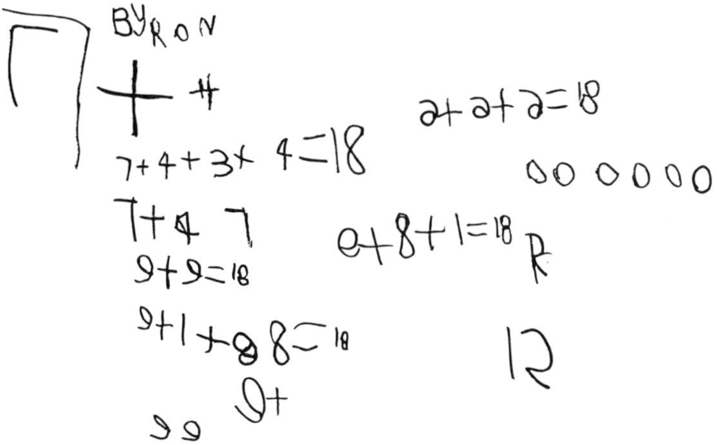

Fig. 16.2. Byron's ways of making 18 by using his calculator

The different ways the children approached the teddy bear activity illustrate their different levels of understanding of addition. Most of these kindergarten children did not yet understand that addition is an efficient alternative to recounting. Calculator use, by separating the choice of operation from the means of carrying out the computation, can assist children in understanding the purpose and use of the four operations.

The following two examples illustrate how calculators were used to link real partitioning situations with the division operation and also to begin the long-term development of an understanding of decimals. Grade 2 children who had carried out a tree survey were making a pictograph of their results. One group of seven children needed to cut out sixty-four pictures of trees to paste onto the chart. The teacher asked how many trees each child in the group would need to contribute. Using their calculators, some children divided 64 by 7 and obtained 9.1428571. The teacher asked what the answer meant. A child quickly replied, "It is nine and a bit. So if we made ten each we would have some left over—actually we would have six left over." The children saw that division was an appropriate operation to use to answer this question, and the connection between the real problem and the calculation provided a context in which sense could be made of the unexpectedly long answer. Without calculators, this opportunity to develop an understanding of decimals and division would have been lost because of the difficulty of the calculation.

In the other example, kindergarten children wanted to share fifty-five cookies among their ten teddy bears. One child had discovered the "sharing sign" on the calculator and had informed classmates of her discovery. Two children got the answer 5.5 on their calculator. The teacher discussed the idea that the ".5" on the display represented a half, and one child commented, "Oh, that's five and a half cookies then." Some days later, she explained it to a project visitor, along with the calculation $55 \div 10 = 5.5$. Figure 16.3 shows the ten bears that Zoe and Julienne pasted onto their paper. They have drawn five cookies and a fraction of a cookie alongside each bear.

At some stage during elementary school, children should become adept at mentally dividing and multiplying by 10 and its powers. We would not wish them to be dependent on a calculator for this simple computation. However, it is essential that they do not merely "shift the decimal points around" but know what the process involves and what their answers mean. Using a calculator to carry out divisions in real sharing situations and seeing the answer several years before they can carry out the computation themselves can provide a good basis for this understanding.

Further examples of the ways in which children in the Calculators in Primary Mathematics project used a variety of mental and calculator strategies to solve real problems are recorded on the videotape *Young Children Using Calculators* (Groves and Cheeseman 1993). In one episode, second-grade children, reviewing sporting statistics gathered earlier, were asked, "Sarah's team scored

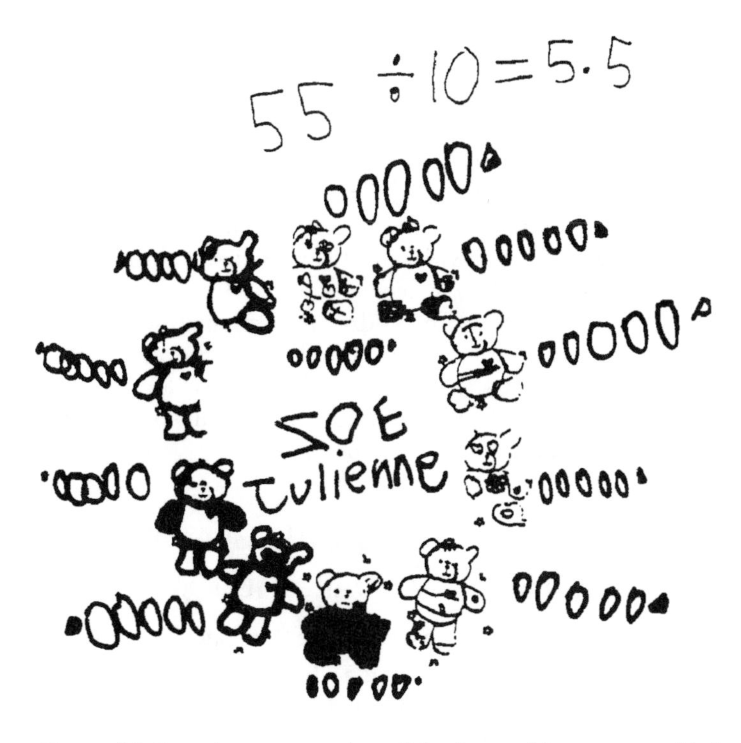

Fig. 16.3. Zoe and Julienne's representation of sharing cookies among teddy bears

49. Jason's team scored 63. How many more runs did Jason's team make?" The children used a wide range of methods to solve the problem. For example, Penny and Tennelle used their fingers to help count on from 49 to 63. Jason explained his method thus: "I worked out that 49 plus 11 was 60. Then I added on another 3 and that was 14." No child used subtraction.

This example illustrates an important point about computational strategies in mathematics. Addition can be done by counting—at least for whole numbers—and children will initially use strategies based on counting by ones. Subtraction can be done by counting on (as the girls did) or by trial addition (as Jason did). Teachers need to encourage children to use whatever methods they have available to solve problems and to use their number sense to choose strategies appropriate to both the situation and the numbers involved. However, in the long term, teachers must lead children to recognize the situations in which each of the four operations can appropriately be used—regardless of how the computations are carried out.

Another problem stated that "everyone in the grade [26 children] did 10 step-ups and 10 star jumps. How many were done all together?" Some children counted around the room by tens. Jason, however, explained, "I knew that 20 tens is 200 and 6 tens is 60. So I knew that will be 260, and I knew that they both will have the same answer because they are the same sum." The teacher asked whether he had worked out how many all together. Jason replied that he hadn't originally done so, but "while I was sitting on the mat, I worked it out in my head." His explanation was "I knew that 200 + 200 is 400. Then 6 + 6 is 12, and 12 tens is 120. So it would be 520."

The teacher commented that before the students used calculators, she would have posed similar problems, but with much smaller numbers. The calculator allowed the children to deal with realistic numbers, although it is interesting that very few children used their calculator to find the answers in a straightforward way. It takes a long time for children to recognize the appropriate operation to use to find the solutions to real problems. In the testing program of the Calculators in Primary Mathematics project, there was some evidence that the children with long-term experience in the use of calculators were better able to make such choices than those who hadn't had such experience.

THE EFFECT OF USING CALCULATORS BEFORE TEACHING STANDARD ALGORITHMS

In order to investigate the long-term effect of calculator use on children's learning of number concepts and skills, an extensive program of testing and interviews was conducted. The performance of third- and fourth-grade children in the Calculators in Primary Mathematics project was compared with that of a control group consisting of the last cohort of third- and fourth-grade children at the same schools who had not been part of the project. A written test, a test of calculator use, and two different interviews were used over the three-year period 1991 to 1993 at grades 3 and 4. In each year, approximately five hundred children completed the written test and the test of calculator use, with more than 10 percent of these children also taking part in one of the two interviews.

Despite fears initially expressed by some parents, there was no evidence that children became reliant on calculators at the expense of their ability to use other methods of computation. The interviews showed that children with long-term experience with calculators performed better overall than children without such experience—both on questions for which they could use any tool of their choice and on mental computation. These children also performed better on a wide range of items involving negative numbers, place value in large numbers, and, most markedly, decimals. They also made more appropriate choices of calculating devices (e.g., calculator, base-ten arithmetic blocks) and were better able to interpret their answers when using a calculator, particularly when decimal answers were involved. No detrimental effects of calculator use were observed in either the interviews or the written tests. For further details of the results from these tests and interviews, see Groves (1993, 1994) and Stacey (1994).

CONCLUSION

Our purpose in introducing calculators was not to make the students dependent on them but rather to enhance that elusive quality "number sense" by providing a rich mathematical environment to explore.

As stated earlier, we believe that the amount of time given to paper-and-pencil algorithms in the school curriculum should be the subject of intense debate and experimentation. Children should be given the power to calculate, but the methods that we expect them to use can change.

Activities like those described above illustrate the tremendous potential calculators have in developing children's conceptual understanding and mental-computation strategies before the formal teaching of relevant algorithms. We maintain that such activities, used over long periods of time, support children's development of number sense.

REFERENCES

Groves, Susie. "The Effect of Calculator Use on Third and Fourth Graders' Computation and Choice of Calculating Device." In *Proceedings of the Eighteenth International Conference for the Psychology of Mathematics Education,* vol. 3, edited by João Pedro da Ponte and João Filipe Matos, pp. 33–40. Lisbon, Portugal: University of Lisbon, 1994.

———. "The Effect of Calculator Use on Third Graders' Solutions of Real World Division and Multiplication Problems." In *Proceedings of the Seventeenth International Conference for the Psychology of Mathematics Education,* vol. 2, edited by Ichiei Hirabayashi, Nobuhiko Nohda, Keiichi Shigematsu, and Fou-Lai Lin, pp. 9–16. Tsukuba, Ibaraki, Japan: University of Tsukuba, 1993.

Groves, Susie, and Jill Cheeseman. *Young Children Using Calculators.* Videotape. Burwood, Victoria, Australia: Video Production Unit, Deakin University, 1993.

Groves, Susie, Jill Cheeseman, Joy Dale, and Ann Dornau. *Calculators in Primary Mathematics: The Newsletters.* Geelong, Victoria, Australia: Centre for Studies in Mathematics, Science and Environmental Education, Deakin University, 1994.

National Research Council. *Reshaping School Mathematics: A Philosophy and Framework for Curriculum.* Washington, D.C.: National Academy Press, 1990.

Shuard, Hilary. "CAN: Calculator Use in the Primary Grades in England and Wales." In *Calculators in Mathematics Education,* 1992 Yearbook of the Natiaonal Council of Teachers of Mathematics, edited by James T. Fey, pp. 33–45. Reston, Va.: National Council of Teachers of Mathematics, 1992.

Stacey, Kaye. "Arithmetic with a Calculator: What Do Children Need to Learn?" In *Challenges in Mathematics Education: Constraints on Construction,* edited by Garry Bell, Bob Wright, Neville Leeson, and John Geake, pp. 563–70. Lismore, New South Wales, Australia: Mathematics Education Research Group of Australasia, 1994.

17

The Harmful Effects of Algorithms in Grades 1–4

Constance Kamii

Ann Dominick

STARTING in the 1970s, researchers such as Ashlock (1972, 1976, 1982) and Brown and Burton (1978) have documented the erroneous but consistent ways in which children inadvertently change the algorithms for multidigit computation. The rules children made up showed that their focus was on trying to remember the steps instead of on logically solving the problems.

Although some researchers studied children's unsuccessful efforts to use the conventional algorithms, others reported the surprising procedures invented by many children, adolescents, and adults. (Throughout this paper, the term *algorithm* is used to refer to the conventional rules of "carrying," "borrowing," etc.; child-invented procedures are referred to as *procedures.*) Cochran, Barson, and Davis (1970), for example, described how an eight-year-old solved 62 – 28: first, 60 – 20 = 40, then, 2 – 8 = –6, and finally, 40 – 6 = 34. Similar findings have been reported in Argentina (Ferreiro 1988, personal communication 1976), the Netherlands (ter Heege 1978), England (Plunkett 1979), and South Africa (Murray and Olivier 1989).

By the 1980s, some researchers were seriously questioning the wisdom of teaching conventional algorithms. In Brazil, Carraher, Carraher, and Schliemann (1985) and Carraher and Schliemann (1985) compared children who used algorithms with those who used their own procedures. Posing as a customer, for example, the researcher asked a child street vendor how much four coconuts would cost if one cost 35 cruzeiros. The vendor replied, "Three will be 105, plus ... 35 ... 140!" (Carraher, Carraher, and Schliemann 1985, p. 26). In a subsequent interview, however, the same child wrote the answer 200, as shown in figure 17.1, and explained, "Four times 5 is 20, carry the 2; 2 plus 3 is 5, times 4 is 20" (p. 26). The researchers concluded that children who use their own procedures are much more likely to produce correct answers than those who try to use algorithms. They thus began to think that algorithms were a hindrance

```
  2
  35
× 4
200
```

Fig. 17.1. A Brazilian child's way of using the algorithm

rather than a help. Jones (1975) in England, Vakali (1984) in Greece, and Dominick (1991) in the United States reached the same conclusion.

Some investigators went further in the 1990s and concluded that algorithms are harmful to children. Narode, Board, and Davenport (1993) compared second graders before and after they were taught algorithms and concluded that children lose conceptual knowledge when they learn these rules. Kamii (1994) compared children in grades 2–4 who had been taught algorithms with those who had never been taught any algorithms and found that those who did their own thinking got more correct answers and had much better knowledge of place value. She also pointed out that the algorithms that are now conventional are the results of centuries of construction by adult mathematicians. Although it is not necessary for children to go through every historical step, it is unrealistic to expect them to skip the entire process of construction.

Many algorithms that were conventional centuries ago reveal a parallel between an individual's construction of numerical reasoning and humanity's construction of these rules. For example, some Hindus added 278 and 356 on a "dust" board in the following way (Groza 1968):

```
278 ------- 578 ------- 628 ------- 634
356         56          6
```

In this algorithm, 200 and 300 of the 278 and 356 were added first and erased, and changed to 500 (the "5" of 578). The next step was to add 70 and 50, erase them and the 500, and change them to 620 (the "62" of 628). The 8 and 6 were then added and erased, as well as the 2, and changed to 34.

Some leaders in mathematics education also began to say that we must stop teaching algorithms because they make no sense to most children and discourage logical thinking. The most convincing arguments based on systematic study of children in classrooms were advanced by Madell (1985), Burns (1994), and Leinwand (1994).

The purpose of this paper is to present the evidence that led us to the conviction that algorithms not only are not helpful in learning arithmetic but also hinder children's development of numerical reasoning. We begin by discussing our data and go on to describe teachers' observations.

RESEARCH BASED ON PIAGET'S CONSTRUCTIVISM

The distinction Piaget made among the three kinds of knowledge on the basis of their ultimate sources shows why teaching conventional algorithms does not foster children's learning of mathematics. The three kinds of knowledge he distinguished are physical, social, and logico-mathematical knowledge.

Physical knowledge is knowledge of objects in external reality. The color and weight of a block are examples of physical properties that are *in* objects in external reality and can be known empirically by observation.

Examples of *social (conventional) knowledge* are holidays, such as the Fourth of July, and written and spoken languages, such as the word *block.* Whereas an ultimate source of physical knowledge is in objects, an ultimate source of social knowledge is in conventions made by people.

Logico-mathematical knowledge consists of mental relationships, and the ultimate source of these relationships is each person's mental actions. For example, the child's knowing that one quantity combined with another gives a larger quantity results from his or her making a mental relationship. Someone else can explain this relationship, but this explanation does not become the child's knowledge until he or she makes the relationship. Likewise, an adult can explain to a child the algorithm for two-digit addition. However, listening to this explanation does not ensure the child's making the mental relationships about how to combine the two quantities.

A characteristic of logico-mathematical knowledge is that there is nothing arbitrary in it. For example, adding 356 to 278 results in 634 in every culture. The social (conventional) rule, or algorithm, stating that one *must* add the ones first, then the tens, and then the hundreds is arbitrary. The teaching of algorithms is based on the erroneous assumption that mathematics is a cultural heritage that must be *transmitted* to the next generation.

Piaget's constructivism and the more than sixty years of scientific research by him and others all over the world led Kamii to a compelling hypothesis: Children in the primary grades should be able to invent their own arithmetic without the instruction they are now receiving from textbooks and workbooks. This hypothesis was amply verified, as can be seen in Kamii (1985, 1989b, 1994).

A significant byproduct of this research was the finding that when children are encouraged to do their own thinking to add, subtract, and multiply multidigit numbers, they always deal with the large units first, such as the tens, and then with the ones. As can be seen in figure 17.2 and Kamii (1989a, 1989b, 1994), this finding confirmed Madell's (1985) statement that when children are encouraged to think in their own ways, they "*universally* proceed from left to right" (p. 21).

Another significant finding was that at the end of second and third grade, the children in "constructivist" classes consistently excel over those in traditional classrooms where algorithms are taught. Hypothesizing that algorithms are harmful to children, Kamii compared the performance of children who had never been taught these conventional rules with those who had.

Problem			
18 +17	10 + 10 = 20 8 + 7 = 15 20 + 10 = 30 30 + 5 = 35	10 + 10 = 20 8 + 2 = another ten 20 + 10 = 30 30 + 5 = 35	10 + 10 = 20 7 + 7 = 14 14 + 1 = 15 20 + 10 = 30 30 + 5 = 35
44 −15	40 − 10 = 30 4 − 5 = 1 less than 0 30 − 1 = 29	40 − 10 = 30 30 − 5 = 25 25 + 4 = 29	40 − 10 = 30 30 + 4 = 34 34 − 5 = 29
135 × 4	4 × 100 = 400 4 × 30 = 120 4 × 5 = 20 400 + 120 + 20 = 540	4 × 100 = 400 4 × 35 = 70 + 70 = 140 400 + 140 = 540	

Fig. 17.2. Procedures invented by children for addition, subtraction, and multiplication

In the school where she worked in 1989–91, some teachers taught algorithms whereas others did not, according to the following distribution:

Grade 1: None of the four teachers taught algorithms.

Grade 2: One of the three teachers taught algorithms; of the two who did not, one convinced parents that they should not teach algorithms at home, either.

Grade 3: Two of the three teachers taught algorithms.

Grade 4: All four teachers taught algorithms.

All the classes were heterogenous and comparable (the principal mixed up all the children at each grade level and divided them as randomly as possible before each school year). Students who transferred in from other schools were also distributed randomly among all the classes.

One of the problems Kamii asked each child to solve mentally in individual interviews was 7 + 52 + 186 (or 6 + 53 + 185). The answers given by the second, third, and fourth graders are presented in tables 17.1–17.3. It can be seen in tables 17.1 and 17.2 that the "No algorithms" classes, both in second and in third grade, produced the highest percentages of correct answers (45 percent and 50 percent respectively). It is also evident that the "No algorithms" second- and third-grade classes produced more correct answers than all the fourth-grade classes (table 17.3), who were all taught algorithms.

More significant are the incorrect answers listed in tables 17.1–17.3. All the wrong answers given in each class appear in these tables. The broken lines through the middle of each table indicate a range of answers that can be considered reasonable. It can be seen in table 17.1 that the incorrect answers of the "No algorithms" class were much more reasonable than the wrong answers of the "Algorithms" class. (The class that was exposed to some algorithms at home came out in between.) In third grade (table 17.2), too, the incorrect answers of the "No algorithms" class were much more reasonable than those of the "Algorithms" classes. The fourth graders (table 17.3), who had had an additional year of algorithms, gave incorrect answers that were more unreasonable than those of the third-grade "Algorithms" classes. In fourth grade, there were more answers in the 700s and 800s, and some four- and five-digit answers. A new symptom also emerged in fourth grade: Answers such as "4, 4, 4" consisting of single digits, demonstrating that children were thinking about three independent columns.

TABLE 17.1
Answers to 7 + 52 + 186 Given by Three Classes of Second Graders in May 1990

Algorithms $n = 17$	Some algorithms taught at home $n = 19$	No algorithms $n = 20$
9308		
1000		
989		
986		
938	989	
906	938	
838	810	
295	356	617
		255
		246
245 (12%)	245 (26%)	245 (45%)
		243
		236
		235
200	213	138
198	213	—
30	199	—
29	133	—
29	125	—
—	114	
—	—	
	—	
	—	

Note. Blanks indicate that the child declined to try to work the problem.

TABLE 17.2
Answers to 6 + 53 + 185 Given by Three Classes of Third Graders in May 1991

Algorithms $n = 19$	Algorithms $n = 20$	No algorithms $n = 10$
	800 + 38	
838	800	
768	444	
533	344	284
246		245
244 (32%)	244 (20%)	244 (50%)
235	243	243
234	239	238
	238	
	234	
213	204	221
194	202	
194	190	
74	187	
29	144	
—	139	
—	—	
	—	

Note. Blanks indicate that the child declined to try to work the problem.

Why Algorithms Are Harmful

We have two reasons for saying that algorithms are harmful: (1) They encourage children to give up their own thinking, and (2) they "unteach" place value, thereby preventing children from developing number sense.

As stated earlier, when children invent their own procedures, they proceed from left to right. Because there is no compromise possible between going toward the right and going toward the left as the algorithms require, children have to give up their own thinking to use the algorithms.

When we listen to children using the algorithm to do

$$\begin{array}{r} 89 \\ \underline{+34} \end{array},$$

for example, we can hear them say, "Nine and four is thirteen. Put down the three; carry the one. One and eight is nine, plus three is twelve...." The algorithm is convenient for adults, who already know that the "one," the "eight," and the "three" stand for 10, 80, and 30. However, for primary school children, who have a tendency to think that the "8" means eight, and so on, the algorithm serves to reinforce this error. The incorrect answers given by the "Algorithms" classes in tables 17.1–17.3 demonstrate that the algorithms

TABLE 17.3
Answers to 6 + 53 + 185 Given by Four Classes of Fourth Graders in May 1991

Algorithms $n = 20$	Algorithms $n = 21$	Algorithms $n = 21$	Algorithms $n = 18$
	1215		
	848		
	844		
	783		
1300	783		10,099
814	783		838
744	718	791	835
715	713	738	745
713 + 8	445	721	274
	245		
244 (30%)	244 (24%)	244 (19%)	244 (17%)
243	234		234
	224		234
			234
194	194	144	225
177	127	138	“8, 3, 8”
144	—	134	“4, 3, 2”
143	—	“8, 3, 7”	“4, 3, 2”
134		“8, 1, 7”	—
“4, 4, 4”		—	—
“1, 3, 2”		—	—
		—	
		—	
		—	
		—	
		—	
		—	

Note. Blanks indicate that the child declined to try to work the problem.

“untaught” place value and prevented the children from developing number sense. The children in all the “Algorithms” classes did not notice that their answers of 144, 783, and so on, were unreasonable for 6 + 53 + 185.

Most of the children in the “No algorithms” classes typically began by saying, “One hundred eighty and fifty is two hundred thirty.” This is why their errors were reasonable even when they got an incorrect answer. Those in the “Algorithms” classes, however, typically said, “Six and three is nine, plus five is fourteen. Put down the four; carry the one.…” Many of them then added 6 (the first addend) to the 1 in 185 (the third addend) and got an answer in the 700s or 800s.

Observations in Classrooms

The harmful effects of algorithms became even more evident when, in 1991–92, one of the fourth-grade teachers, Cheryl Ingram, decided to change her teaching to a constructivist approach. One of the ways in which she tried to wean the children away from algorithms was to write problems such as 876 + 359 horizontally on the board and to ask the class to invent many different ways of solving them without using a pencil. As the children volunteered to explain how they got the answer of 1235 by using the algorithm in their heads, she wrote exactly what students said for each column (6 + 9 = $\underline{15}$, 7 + 5 + 1 = $\underline{13}$, and 8 + 3 + 1 = $\underline{12}$) as follows:

$$\begin{array}{r} 15 \\ 13 \\ \underline{+12} \\ 40 \end{array}$$

After the child finished explaining how he or she got the answer of 1235, Ms. Ingram said, "But I followed your way, and when I put 15, 13, and 12 together, I got 40 as my answer. How did you get 1235?" Most of the children were stumped and became silent, until someone pointed out that the teacher's 13 was really 130 and that her 12 stood for 1200.

This kind of place-value problem was not too hard to correct. The persisting difficulty lay in the column-by-column, single-digit approach that prevented children from thinking about multidigit numbers. Presented with problems like 876 + 359, the children continued to give fragmented answers from right to left, such as "5, 130, 1200" (for 6 + 9, 10 + 70 + 50, and 100 + 800 + 300, respectively).

One day, to encourage the children to think about multidigit numbers, Ms. Ingram put on the board one problem after another that had 99 (or 98 or 95) in one of the addends, such as 366 + 199, 493 + 99, and 601 + 199. Only this kind of problem was presented during the entire hour, and the children were asked as usual to think about different ways of adding them.

Almost all the children in the class continued to use the algorithm during the entire hour. One child, however, whom we will call Joe, had been in "constructivist" classes since first grade and volunteered solutions like the following for each problem: "I changed '366 + 199' to '365 + 200,' and my answer is 565." After an entire hour of this kind of "interaction," the number of children imitating Joe by the end of the hour was only three! The rest of the class continued to deal with each column separately.

The school year proceeded with many ups and downs as Ms. Ingram continued her struggle to revive the children's own thinking (see Kamii [1994] for further detail). In May 1992, 6 + 53 + 185 was presented to her fourth graders, and the results were gratifying, as can be seen in figure 17.3.

In figure 17.3, the top row of each 2 × 2 matrix shows the number of children who gave the correct answer, and the bottom row shows the number of

1991

	Algorithm	Invented Procedures
Correct Answer	3	0
Incorrect Answers	13	1

1992

	Algorithm	Invented Procedures
Correct Answer	0	15
Incorrect Answers	2	3

(One child was excluded from this analysis because she said she was thinking of multiplying 185 by 53 and of adding 6.)

Fig. 17.3. Fourth graders' use of the algorithm and invented procedures and the correctness of their answer to 6 + 53 + 185 in May 1991 and May 1992

those who gave incorrect answers. The left-hand column of each matrix indicates the number who used the conventional algorithm, and the right-hand column shows the number who used their own invented procedures. By comparing these matrices, we can see that when Ms. Ingram taught algorithms in 1990–91, almost all her students used the algorithm, and most of them got incorrect answers (shown in the last column of table 17.3). In 1991–92, by contrast, when Ms. Ingram encouraged her students to do their own thinking, most of her students used invented procedures and got the correct answer.

Ann Dominick, the second author, is a classroom teacher who has taught third and fourth graders for twelve years. When she worked with fourth graders in one school, almost every child in her class had been taught algorithms before coming to her. Now that she teaches third graders in another school, most of the children in her class have *never* been taught these algorithms. The difference in students' thinking is astounding.

The most striking differences are in students' confidence and their knowledge of place value. Those who have made sense of mathematics approach it with confidence rather than fear and hesitation. The students' intellectual pace is a gallop instead of a walk.

At the beginning of each year Ms. Dominick conducts individual interviews with each student to assess, among other things, their knowledge of place value. In the place-value task (Kamii 1989b), children are asked to show with sixteen counters what each digit in the numeral 16 means. When students came to her class using algorithms, approximately 20 percent of the fourth graders each year showed ten counters for the 1 in 16. (The other 80 percent showed only one counter.) In the third-grade classes where most of the children have not been taught any algorithms, about 85 percent show ten counters for the 1 in 16.

Ms. Dominick's thinking about teaching algorithms has changed over the years from (*a*) teaching arithmetic *by* teaching algorithms to (*b*) teaching the algorithms after "laying the groundwork for understanding" to (*c*) not teaching algorithms at all. The final shift came from reflecting on what happened when she "laid the groundwork for understanding" and then taught the algorithm.

The first rationale for teaching the algorithms was that it seemed to be the most efficient method. Once students started inventing their own methods, however, this argument no longer held true. For example, using the algorithm for multiplying by 25 often *slows* students' thinking. Frequently, children use their knowledge of $25 \times 4 = 100$ to reason that $25 \times 16 = 400$. Similarly, it takes much more time to use the algorithm to compute $502 - 304$. A more efficient way is to reason that $500 - 300 = 200$ and that $200 - 2 = 198$. An understanding of place value and reference points such as $25 \times 4 = 100$ and $250 \times 4 = 1000$ allow children the flexibility to determine for themselves the most efficient method for solving a problem in a given situation.

The second argument for teaching algorithms was to give struggling students a method for getting answers. It seemed that these students deserved to be given a method for at least getting an answer to achieve some degree of success. It later became evident, however, that when these students forgot a step or developed a "buggy" algorithm, they had nothing to fall back on. Teaching algorithms to these students also sent them the message that "the logic of this procedure is too much for you; so just follow these steps and you'll get the right answer." Some students need more time than others to develop the logic of mathematics. These children deserve the time they need to develop confidence in their ability to make sense of mathematics.

CONCLUSION

Children come to school with enormous potential for powerful thinking. Educators must try to develop this potential instead of continuing to "put the cart in front of the horse." Adults may pack the cart with treasures, but children need to go through their own constructive process and to proceed with confidence in their own ability to solve problems every step of the way.

REFERENCES

Ashlock, Robert B. *Error Patterns in Computation.* Columbus, Ohio: Charles E. Merrill Publishing Co., 1972, 1976, 1982.

Brown, John Seely, and Richard R. Burton. "Diagnostic Models for Procedural Bugs in Basic Mathematical Skills." *Cognitive Science* 2 (1978): 155–92.

Burns, Marilyn. "Arithmetic: The Last Holdout." *Phi Delta Kappan* 75 (1994): 471–76.

Carraher, Terezinha Nunes, David William Carraher, and Analucia Dias Schliemann. "Mathematics in the Streets and in Schools." *British Journal of Developmental Psychology* 3 (1985): 21–29.

Carraher, Terezinha Nunes, and Analucia Dias Schliemann. "Computation Routines Prescribed by Schools: Help or Hindrance?" *Journal for Research in Mathematics Education* 16 (1985): 37–44.

Cochran, Beryl S., Alan Barson, and Robert B. Davis. "Child-Created Mathematics." *Arithmetic Teacher* 17 (March 1970): 211–15.

Dominick, Ann McNamee. "Third Graders' Understanding of the Multidigit Subtraction Algorithm." Doctoral dissertation, Peabody College for Teachers, Vanderbilt University, 1991.

Ferreiro, Emilia. *Alfabetizacão em processo.* São Paulo: Cortez,1988.

Groza, Vivian S. *A Survey of Mathematics: Elementary Concepts and Their Historical Development.* New York: Holt, Rinehart & Winston, 1968.

Jones, D. A. "Don't Just Mark the Answer—Have a Look at the Method!" *Mathematics in School* 4 (May 1975): 29–31.

Kamii, Constance. *Double-Digit Addition: A Teacher Uses Piaget's Theory.* Videotape. New York: Teachers College Press, 1989a.

———. *Young Children Continue to Reinvent Arithmetic, 2nd Grade.* New York: Teachers College Press, 1989b.

———. *Young Children Continue to Reinvent Arithmetic, 3rd Grade.* New York: Teachers College Press, 1994.

———. *Young Children Reinvent Arithmetic.* New York: Teachers College Press, 1985.

Leinwand, Steven. "It's Time to Abandon Computational Algorithms." *Education Week*, 9 February 1994, p. 36.

Madell, Rob. "Children's Natural Processes." *Arithmetic Teacher* 32 (March 1985): 20–22.

Murray, Hanlie, and Alwyn Olivier. "A Model of Understanding Two-Digit Numeration and Computation." In *Proceedings of the Thirteenth Meeting of the International Group for the Psychology of Mathematics Education,* edited by Gerard Vergnaud, Janine Rogalski, and Michele Artigue, pp. 3–10. Paris: Laboratoire PSYDEE of the National Center of Scientific Research, 1989.

Narode, Ronald, Jill Board, and Linda Davenport. "Algorithms Supplant Understanding: Case Studies of Primary Students' Strategies for Double-Digit Addition and Subtraction." In *Proceedings of the Fifteenth Annual Meeting, North American Chapter of the International Group for the Psychology of Mathematics Education,* Vol. 1, edited by Joanne Rossi Becker and Barbara J. Pence, pp. 254–60. San Jose, Calif.: San Jose State University, Center for Mathematics and Computer Science Education, 1993.

Plunkett, Stuart. "Decomposition and All That Rot." *Mathematics in School* 8, no. 3 (1979): 2–7.

ter Heege, Hans. "Testing the Maturity for Learning the Algorithm of Multiplication." *Educational Studies in Mathematics* 9 (1978): 75–83.

Vakali, Mary. "Children's Thinking in Arithmetic Word Problem Solving." *Journal of Experimental Education* 53 (1984): 106–13.

18

A Contextual Investigation of Three-Digit Addition and Subtraction

Kay McClain

Paul Cobb

Janet Bowers

DISCUSSIONS of the role of algorithms in the elementary school curriculum give rise to conflicting notions of how algorithms should be approached in the classroom. These views range from encouraging students to invent their own algorithms with minimal guidance to teaching students to perform traditional algorithms. In this paper, we will discuss an approach that eschews both these extremes. This approach values students' construction of non-standard algorithms. However, it also emphasizes the essential role of the teacher and of instructional activities in supporting the development of students' numerical reasoning. In addition, this approach highlights the importance of discussions in which students justify their algorithms. It therefore treats students' development of increasingly sophisticated algorithms as a means for conceptual learning. In our view, an approach of this type is consistent with reform recommendations, like those of the National Council of Teachers of Mathematics (1989), that stress the need for students to develop what Skemp (1976) calls a relational understanding rather than merely to memorize the steps of standard procedures. The contrast between this

The research and analysis reported in this paper was supported by the Office of Educational Research and Improvement (OERI) under grant number R305A60007 through the National Center for Improving Student Learning and achievement in Mathematics and Science. The opinions expressed do not necessarily reflect the views of OERI.

approach and traditional instruction is captured by the difference between "How can I figure this out?" and "What was I told I was supposed to do?" (Yackel et al. 1990). In the approach we will discuss, relational understanding occurs as teachers systematically support students' construction of personally meaningful algorithms. These algorithms emerge when students engage in sequences of problem-solving activities designed to provide opportunities for them to make sense of their mathematical activity. An immediate consequence of this focus on algorithms as symbolizing numerical relationships is that it precludes the unreasonable answers that often occur from the misapplication of little-understood standard algorithms.

To clarify our viewpoint, we will present episodes taken from a third-grade classroom in which we conducted a nine-week teaching experiment. (The authors of this paper were all members of the research team involved in the teaching experiment. The first author was also the classroom teacher who appears in the episodes in this paper.) One of the goals of the experiment was to develop an instructional sequence designed to support third graders' construction of increasingly sophisticated conceptions of place-value numeration and increasingly efficient algorithms for adding and subtracting three-digit numbers. Our intent is not to offer examples of exemplary teaching but instead to provide a context in which to examine the relationship between numerical understanding and the use of powerful algorithms. In addition, the episodes will illustrate how students' beliefs about mathematics in school influence their development of personally meaningful algorithms. In the following sections of this paper, we will first outline the instructional sequence developed in the course of the experiment and then focus on the students' algorithms and their beliefs.

THE INSTRUCTIONAL SEQUENCE

The instructional sequence we designed to support the development of understanding and computational facility in an integrated manner centered on the scenario of a candy factory. The activities initially involved Unifix cubes as substitutes for candies and later involved the development of ways of recording transactions in the factory (cf. Cobb, Yackel, and Wood [1992] and Bowers [1995]). During initial whole-class discussions, the students and teacher negotiated the convention that single pieces of candy were packed into rolls of ten and ten rolls were packed into boxes of one hundred. The ensuing activities included tasks in estimating and quantifying designed to support the development of enumeration strategies. These activities involved showing the students drawings of rolls and pieces with an overhead projector and asking them to determine how many candies there were in all. In addition, the students were shown rectangular arrays of individual candies and asked to estimate how many rolls could be made from the candies shown.

Instructional activities developed later in the sequence included situations in which the students "packed" and "unpacked" Unifix cubes into bars or rolls of ten.

To help students develop a rationale for these activities, the teacher explained that the factory manager, Mr. Strawberry, liked his candies packed so that he could tell quickly how many candies were in the factory storeroom. In order to record their packing and unpacking activity, the students developed drawings and other means of symbolizing as models of their mathematical reasoning (Gravemeijer 1997). The goal of subsequent instructional activities was then to support the students' efforts to mathematize their actual and recorded packing and unpacking activity so that they could interpret it in terms of the composition and decomposition of arithmetical units. To encourage this process, the teacher elaborated on the students' contributions by describing purely numerical explanations in terms of the packing activity and vice versa. This, in turn, served to support the students' development of situation-specific imagery of the transactions in the factory that would provide a foundation for students' mathematical reasoning throughout the sequence.

In a subsequent phase of the sequence, the students were asked to make drawings to show *different ways* that a given number of candies might be arranged in the storeroom if the workers were in the process of packing them. For example, 143 candies might be packed up into one box and four rolls, with three single pieces, or they might be stored as twelve rolls and twenty-three pieces. When the students first described their different ways, they either drew pictures or used tally marks or numerals to record and explain their reasoning. Later, the teacher encouraged all the students to use numerals. To this end, she introduced an inventory form that was used in the factory to keep track of transactions in the storeroom. The form consisted of three columns that were headed from left to right, "Boxes," "Rolls," and "Pieces" (see fig. 18.1). The issue of how to symbolize these different ways and thus the composition and decomposition of arithmetical units became an explicit topic of discussion and a focus of activity. For example, a typical suggestion for verifying that 143 candies could be symbolized as twelve rolls and twenty-three pieces was to pack twenty pieces into two rolls and then pack ten rolls into a box, with three pieces remaining, as shown in figure 18.1.

As a final phase of the sequence, the inventory form was used to present addition and subtraction tasks in what was, for us, the standard vertical-column format (see fig. 18.2). These problems were posed in the context of Mr. Strawberry's filling orders by taking candies from the storeroom and sending them to shops or by increasing his inventory as workers made more candies. The different ways in which the students conceptualized and symbolized these transactions gave rise to discussions that focused on the students' emerging addition and subtraction algorithms.

Boxes	Rolls	Pieces
1	4 14 12	3 23

Fig. 18.1. The inventory form for the candy factory

Boxes	Rolls	Pieces
3	2	7
+ 2	5	8

Fig. 18.2. An addition problem posed on the inventory form

CLASSROOM EPISODES

Throughout the teaching experiment, the students engaged in problem-solving tasks that focused on transactions in the candy factory. As the sequence progressed, we inferred that most of the students' activity was grounded in the situation-specific imagery of the candy factory. Our primary source of evidence was that the students' explanations appeared to have numerical significance in that they spoke of the quantities signified by drawings and numerals instead of merely specifying how they had manipulated digits. Further, they often referred to conventions in the factory such as packing up in the storeroom. Examples of students' varied yet personally meaningful ways of calculating are shown in figure 18.3.

The nonstandard approach of starting the computation with the boxes (the hundreds column in traditional algorithms) (see, e.g., figs. 18.3a and 18.3d) was used by many of the students and became an acceptable way to solve tasks. Since the goal was *not* to ensure that all the students would eventually use the traditional algorithm, the teacher continued to support the development of solutions that could be justified in quantitative terms to other members of the classroom community. Thus, the focus in discussions was on the numerical meanings that the students' records on the inventory form had for them.

Not until the ninth and final week of the teaching experiment did the issue of where to start when calculating emerge as an explicit topic of conversation. The following problem was posed:

> There are five boxes, two rolls, and seven pieces in the storeroom. Mr. Strawberry sends out an order for one box, four rolls, and two pieces. What is left in the storeroom?

Aniquia offered the first solution, shown in figure 18.4. She explained that she first took two pieces from the seven pieces in the storeroom and that she

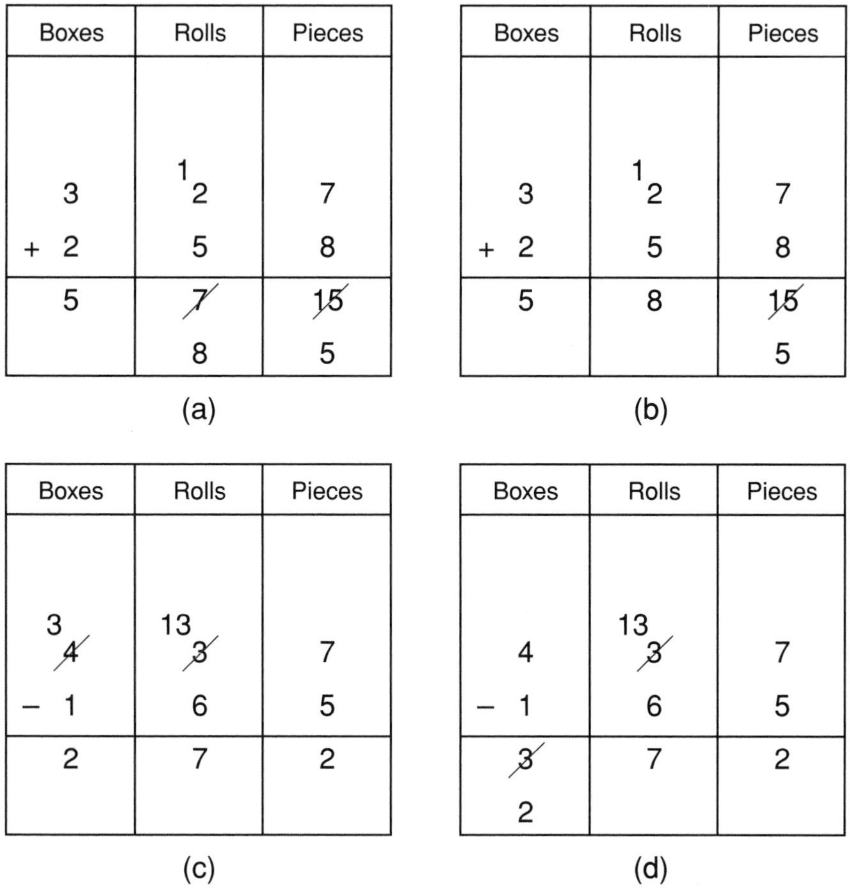

(a)

Boxes	Rolls	Pieces
	1	
3	2	7
+ 2	5	8
5	~~7~~	~~15~~
	8	5

(b)

Boxes	Rolls	Pieces
	1	
3	2	7
+ 2	5	8
5	8	~~15~~
		5

(c)

Boxes	Rolls	Pieces
3	13	
~~4~~	~~3~~	7
− 1	6	5
2	7	2

(d)

Boxes	Rolls	Pieces
	13	
4	~~3~~	7
− 1	6	5
~~3~~	7	2
2		

Fig. 18.3. Students' solution methods to addition and subtraction tasks

then unpacked a box so that she could send out four rolls. The teacher drew pictures of boxes, rolls, and pieces to help other students understand Aniquia's reasoning (see fig. 18.4).

After checking whether the students had any questions for Aniquia, the teacher asked, "Did anybody do it a different way?" Bob raised his hand and shared his solution method, which involved sending out a box first (see fig. 18.5). After Bob had finished his explanation, the teacher asked, "All right. Now, Bob said I'm going to send out my box before I send out my rolls, and Aniquia said I'm going to send out my rolls before I send out my box. Does it matter?"

Some of the students responded:

Cary: It depends on the kind of problem it is.

Avery: Nope, you can do it either way.

Boxes	Rolls	Pieces
4 ~~5~~	12 ~~2~~	7
− 1	4	2
3	8	5

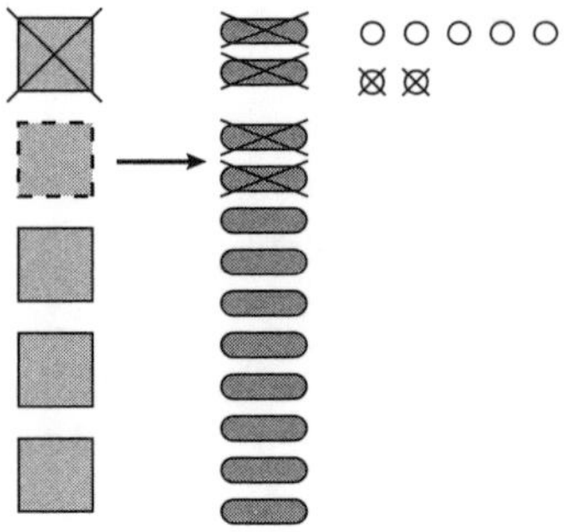

Fig. 18.4. Aniquia's solution on the inventory form and the teacher's graphical explanation

Ann: When you usually do subtraction you always start at the right 'cause the pieces are like the ones.

Avery: But it doesn't matter.

Rick: Yeah, 'cause we're in the candy factory, not in the usual stuff.

At the end of this session, the project team reflected on the students' discussions about where to start and agreed that it would be beneficial to revisit the issue. We thought that many of the students were making the decision on the basis of what they knew about the standard algorithm. The students were third graders, and they had received two years of traditional instruction and had been taught the standard algorithms for adding and subtracting two-digit numbers. It appeared from the students' conversations that they had agreed it did not matter because they were in the candy factory and "not in the usual stuff." We speculated that Cary's comment about depending on the kind of problem referred to whether or not it was posed in the context of the candy factory. For these students there appeared to be two different "maths"—regular school mathematics and the mathematics they did with us in the setting of the candy factory. They were therefore able to justify the difference in where to start because of the different norms and expectations in the two instructional situations. In addition, not only did the students seem to believe that the two "maths" had conflicting norms, but they also appeared to hold different beliefs about what constituted

Boxes	Rolls	Pieces
5	12 ~~2~~	7
− 1	4	2
~~4~~ 3	8	5

Fig. 18.5. Bob's solution on the inventory form

acceptable explanations and justifications in each situation. Explanations in the candy factory focused on imagined numerical transactions in the storeroom and on how they could be expressed on the inventory form. School mathematics entailed explanations of symbol manipulations that need not have quantitative significance.

The next day, the following problem was posed to the class:

> There are three boxes, three rolls, and four pieces of candy in the storeroom. Mr. Strawberry gets an order for two boxes, four rolls, and one piece. How many candies are left in the storeroom after he sends out his order?

The first solution was offered by Martin, who completed the task by first drawing boxes, rolls, and pieces on the white board and then recording the result on the inventory form (see fig. 18.6). It is important to note that his description of his solution process was grounded in the imagery of the candy factory, as evidenced by his explanation, which entailed comments such as "First, I sent out the two boxes."

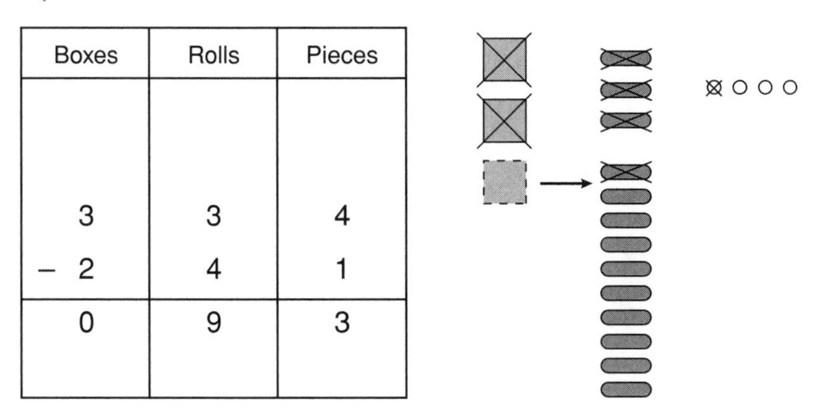

Boxes	Rolls	Pieces
3	3	4
– 2	4	1
0	9	3

Fig. 18.6. Martin's solution and graphical explanation

When Martin had finished, the teacher asked if the students had questions or comments for Martin. Lenny responded that the inventory form did not provide a record of Martin's unpacking actions in the simulated storeroom and that he needed to show that he had unpacked a box. At this point, the teacher decided to take advantage of this reference to notation by reintroducing the question of "where to start." To reflect Lenny's suggestion, she marked Martin's form as though he had first sent out his pieces (see fig. 18.7a). Next, she wrote the same problem on the board again, this time recording a solution in which the boxes were sent out first, as Martin had described (see fig. 18.7b). The following discusion ensued:

Teacher: Does it matter? Is it OK either way?

Rick: No.

Teacher: But we get the same answer.

Rick: But it's confusing. It's harder to work it out. It's easier that way (he points to the form in fig. 18.7a).

Teacher: But my question is, Do we have to do pieces, then rolls, then boxes, or can we do the boxes first?

Bob: No.

Avery: You can start with boxes or rolls.

Rick: Yeah, but that's confusing.

Teacher: OK, let me tell you what I'm hearing you say. I'm hearing you say that it's easier and less confusing [to start with pieces].

Ann: I think you can start with boxes or pieces but not the rolls.

Bob: It would be harder to start with rolls.

Avery: Yeah.

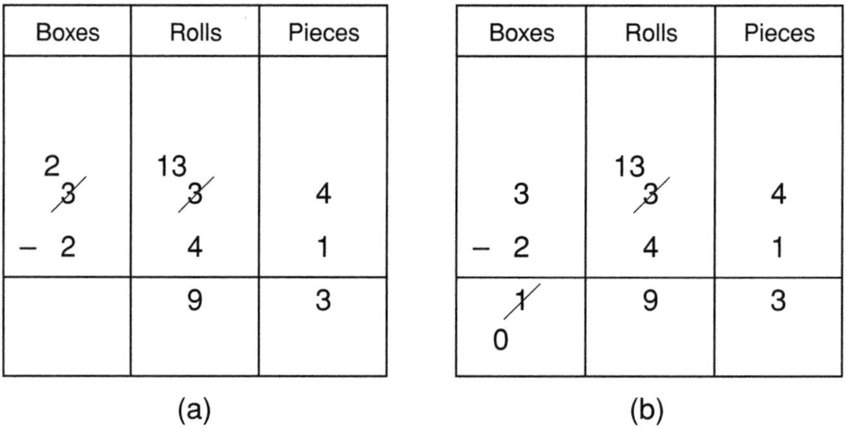

Boxes	Rolls	Pieces
2 ~~3~~ − 2	13 ~~3~~ 4	4 1
	9	3

(a)

Boxes	Rolls	Pieces
3 − 2	13 ~~3~~ 4	4 1
~~1~~ 0	9	3

(b)

Fig. 18.7. Two subtraction solutions

At this point, the discussion shifted to focusing on what was an easier or a less confusing way to solve the problem. The students appeared to agree that both processes were legitimate and that their emphasis was on the ease of comprehension. The teacher's questions were not intended to steer the students to the standard algorithm. Instead, the students' judgments of which approach was easier to understand was made against the background of their prior experiences of doing mathematics in school.

After these discussions, we posed tasks in the traditional column format without the inventory form and discussed the differences in the formats:

Teacher: This is kind of different from the inventory form.

Jess: It's kinda different 'cause it doesn't say boxes, rolls, pieces.

D'Metrius: You still don't have to have boxes, rolls, pieces. It don't have to be up there 'cause you know these are pieces, rolls, and boxes.

Jess: I just know that ... I remember boxes, rolls, and pieces help me.

It appeared from the conversation that many of the students evoked the imagery of the candy factory as they interpreted and solved these tasks. This conjecture was corroborated by the observation that as they worked in pairs on activity sheets, many of the students continued to explain and justify their solutions in terms of activity in the candy factory. In some instances, the students actually drew boxes, rolls, and pieces to support their solutions. The many personally meaningful ways of solving the tasks that emerged are shown in figure 18.8. These included adding from the left (fig. 18.8a), adding from the right (fig. 18.8b), and working from a drawing and recording the result (fig. 18.8c).

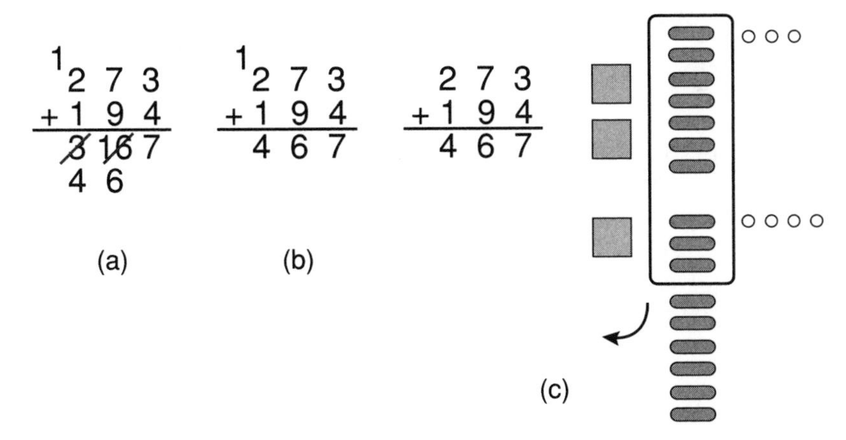

Fig. 18.8. Three different solutions to the problem 273 + 194

CONCLUSION

We have outlined an instructional sequence designed to support students' construction of their own personally meaningful algorithms for three-digit addition and subtraction as it was conducted in one third-grade classroom. Although the situation-specific imagery of the candy factory appeared to support most of the students' activity, the influence of their prior participation in the practices of school mathematics became apparent when they discussed which types of solutions were legitimate and easier to comprehend. However, they did seem to modify their views about what it means to know and do mathematics (at least when interacting with McClain as the teacher) as they compared and contrasted solutions. The important norm that became established was that of explaining and justifying solutions in quantitative terms. We find this significant because the students were able to reorganize their beliefs about doing mathematics in this setting. This change in beliefs could be characterized, using Skemp's (1976) distinction, as shifting from instrumental views toward more relational views of doing mathematics. A comparison of interviews conducted with the students before and after the classroom teaching indicated that they had all made significant progress in their numerical reasoning (Bowers 1996).

The episodes discussed in this paper provide support for a change in emphasis "toward conjecturing, inventing, and problem solving—away from an emphasis on mechanistic answer-finding" (NCTM 1991, p. 3). This emphasis should in no way be construed to mean that students do not need to construct powerful algorithms; it simply calls attention to the importance of students' developing increasingly sophisticated numerical understandings as they develop personally meaningful algorithms. We have illustrated one attempt to achieve this goal that emphasizes the teacher's proactive role, the contribution of carefully sequenced instructional activities, and the importance of discussions in which students explain and justify their thinking.

REFERENCES

Bowers, Janet. "Children's Emerging Conceptions of Place Value in a Technology-Enriched Classroom." Doctoral diss. Peabody College of Vanderbilt University, 1996.

———. "Designing Computer Learning Environments Based on the Theory of Realistic Mathematics Education." In *Proceedings of the Nineteenth Conference of the International Group for the Psychology of Mathematics Education,* edited by Luciano Miera and David Carraher, pp. 202–10. Recife, Brazil: Program Committee of the Nineteenth PME Conference, 1995.

Cobb, Paul, Erna Yackel, and Terry Wood. "A Constructivist Alternative to the Representational View of Mind in Mathematics Education." *Journal for Research in Mathematics Education* 23 (January 1992): 2–33.

Gravemeijer, Koeno. "Mediating between Concrete and Abstract." In *Learning and Teaching Mathematics: An International Perspective,* edited by Terezinha Nunes and Peter Bryant, pp. 315–45. Hove, England: Psychology Press, 1997.

National Council of Teachers of Mathematics. *Curriculum and Evaluation Standards for School Mathematics.* Reston, Va.: National Council of Teachers of Mathematics, 1989.

———. *Professional Standards for Teaching Mathematics.* Reston, Va.: National Council of Teachers of Mathematics, 1991.

Skemp, Richard. "Relational Understanding and Instrumental Understanding." *Mathematics Teaching* 77 (November 1978): 1–7.

Yackel, Erna, Paul Cobb, Terry Wood, Grayson Wheatley, and Graceann Merkel. "The Importance of Social Interaction in Children's Construction of Mathematical Knowledge." In *Teaching and Learning Mathematics in the 1990s,* 1990 Yearbook of the National Council of Teachers of Mathematics, edited by Thomas J. Cooney, pp. 12–21. Reston, Va.: National Council of Teachers of Mathematics, 1990.

19

Children's Invented Algorithms for Multidigit Multiplication Problems

Jae-Meen Baek

A SUBSTANTIAL body of research on children's invented algorithms for solving addition and subtraction problems has shown that children are capable of inventing algorithms to solve various types of real-life problems. When encouraged to invent their own algorithms, students initially model problem situations using concrete materials. As children reflect on and abstract their algorithms, the algorithms become more sophisticated and depend less on the manipulation of concrete materials. Researchers studying students' solutions for multidigit addition and subtraction problems have developed conceptual structures and identified various algorithms that children invent when encouraged to be problem solvers. The research provides evidence that children develop an understanding of addition and subtraction operations through inventing their own algorithms (Fuson et al. 1997).

Understanding multiplication is central to knowing mathematics. The *Curriculum and Evaluation Standards for School Mathematics* (National Council of Teachers of Mathematics 1989) proposed that children need to develop meaning for multiplication by creating algorithms and procedures for operations. What algorithms can or do children invent for multiplication problems? The domain of multiplication is more complex than that of addition or subtraction. Therefore, children may be expected to have more difficulty inventing algorithms for multiplication problems, and their algorithms for them are expected to be more complex. However, this has not been well documented in the literature. This paper discusses the algorithms that students in the upper grades in elementary school invent for multidigit multiplication.

I would like to thank Thomas P. Carpenter for thoughtful suggestions for this paper. I also wish to thank Rebecca Ambrose, Susan B. Empson, Victoria R. Jacobs, Linda W. Levi, and Janet Warfield, who made the observations analyzed in the six classrooms for a full year.

The invented algorithms presented in this paper are based on observations in six classrooms in grades 3–5 in which teachers encouraged students to invent algorithms and to solve problems in different ways. The teachers never taught rules or formal algorithms for children to follow. The students developed their own invented algorithms through sharing different ways of solving problems and discussing the mathematical meaning underlying their inventions.

INVENTED ALGORITHMS FOR MULTIDIGIT MULTIPLICATION PROBLEMS

In the six classes observed, students invented a wide range of algorithms for multiplication problems. Usually, the problems were given to students in word problems and they solved them individually or in small groups. Then the children shared their invented algorithms, discussed similarities or differences among them, and explored underlying mathematical concepts. The students' algorithms were recorded by observers during instruction. The algorithms were classified according to the main schemes that the children used to solve problems.

Children's invented algorithms for multidigit multiplication problems are classified into four categories in this paper: *direct modeling, complete number strategies, partitioning number strategies,* and *compensating strategies.*

A child using the *direct modeling* strategy will model each of the groups using concrete manipulatives or drawings. As the students develop their conceptual understanding of multiplication, their algorithms become more sophisticated. They begin to use *complete number strategies* and *partitioning number strategies.*

Complete number strategies are based on repeated addition of the multiplicands. Children invent more-efficient adding procedures by doubling the multiplicands.

A child using the *partitioning number* strategy will split the multiplicand or multiplier into two or more numbers and create multiple subproblems that are easier to deal with. This procedure allows the children to reduce the complexity of the problem and to use multiplication facts that they already know.

A child using the *compensating strategy* will adjust both multiplicand and multiplier or one of them based on special characteristics of the number combination to make the calculation easier. Children then make corresponding adjustments later if necessary. Each of these algorithms is discussed in the following protocols of students' work observed.

Direct Modeling

Children who use the *direct modeling* strategy model each of the groups using counters, base-ten blocks, tally marks, or other drawings to count the total number of objects. Even with large numbers, some children need to model the entire problem situation and solve it by counting all the objects. There are two types of direct modeling: *direct modeling by ones* and *direct modeling by tens.* The following example is Eric's use of the direct modeling by ones strategy to solve the problem, "If there are 6 classes with 23 children in each class, how many children are there altogether?" Eric drew six rows of twenty-three tally marks and figured out the problem by counting all the tallies by ones (fig. 19.1).

Fig. 19.1. Eric's strategy of direct modeling by ones

Children who use the *direct modeling by tens* strategy model a problem situation with concrete materials using their understanding of base ten and count the materials by tens. The following example is T. J.'s use of the *direct modeling by tens* strategy to solve the problem above. T. J. used blocks of ten to make six groups of twenty-three. He put out two blocks of ten and three unit blocks for each group and then counted the tens and ones blocks separately (see fig. 19.2). The use of direct modeling by tens shows T. J.'s understanding of base ten and allows him to handle a multidigit multiplication problem more efficiently.

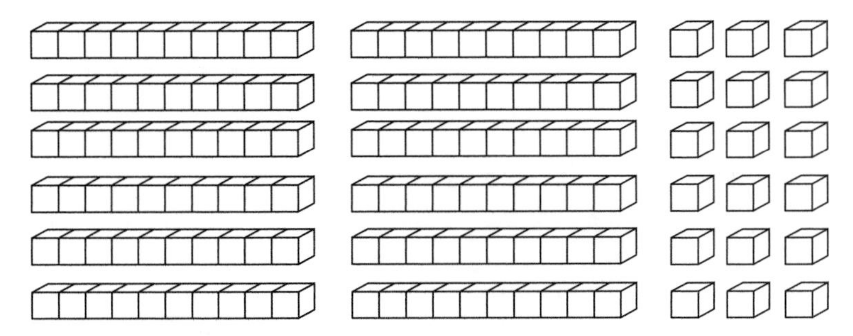

Fig. 19.2. T. J.'s strategy of direct modeling by tens

From their experience modeling problems with concrete materials, children can invent algorithms that are more abstract and not dependent on concrete materials.

Complete Number Strategies

Children who use the *complete number* strategy add the multiplicands but do not partition the multiplier or multiplicand in any particular way. To add the multiplicand, children use several different strategies, such as repeated addition or doubling. The following example illustrates Susie's use of the *complete number strategy using repeated addition* for the same problem. In solving the problem, Susie added 23 six times (fig. 19.3).

Many children use doubling to shorten addition procedures. They make as many pairs of the multiplicand as possible and get a sum for every pair, pairs of pairs, and so on. Tom's used a *complete number strategy using doubling* for the same problem. Tom started by writing down six 23s (see fig. 19.4). By pairing the 23s, he figured out that two 23s is 46. Tom paired the first two 46s and added them to find the sum of 92. Then he added the last 46 to 92 to get 138. Tom wrote down all six 23s but shortened the addition procedure by doubling the multiplicands.

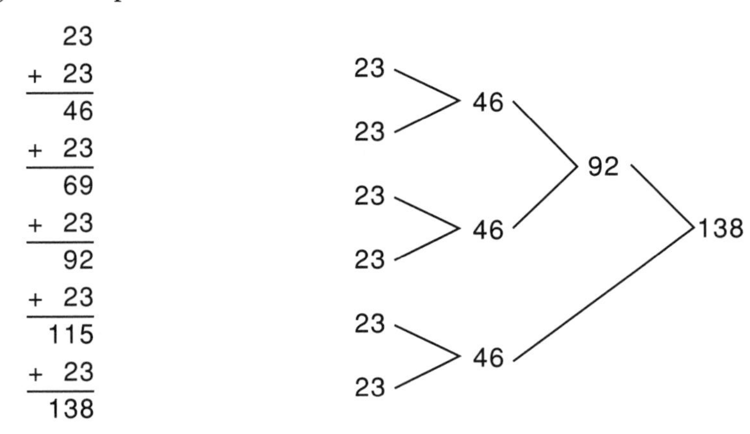

Fig. 19.3. Susie's repeated addition strategy

Fig. 19.4. Tom's doubling strategy

Partitioning Number Strategies

Children who use the *partitioning number* strategy partition the multiplier, the multiplicand, or both into smaller numbers so that they can multiply them more easily. Children partition the multiplier or multiplicand in two different ways: some children partition the numbers into nondecade numbers, and others partition them into decade numbers.

Partitioning a number into nondecade numbers

Children who use this algorithm partition the multiplier or multiplicand into nondecade numbers to make the multiplication process easier or to use multiplication facts that they already know. Figure 19.5 shows Joan's use of

"Three times 177 is ..."
(then writes the following):

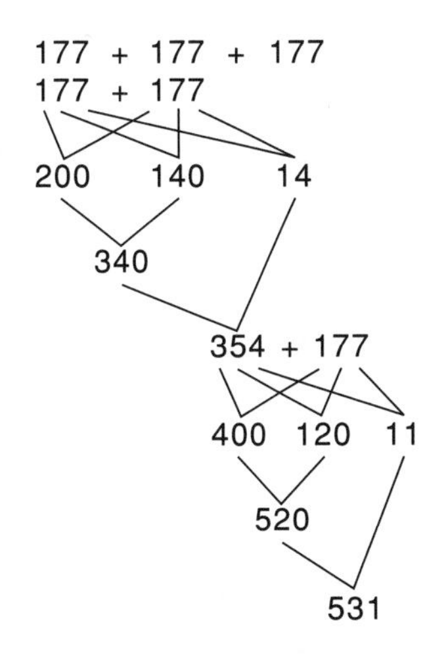

"Five times 531 is ..., hmm.
I can do four times 531 first."

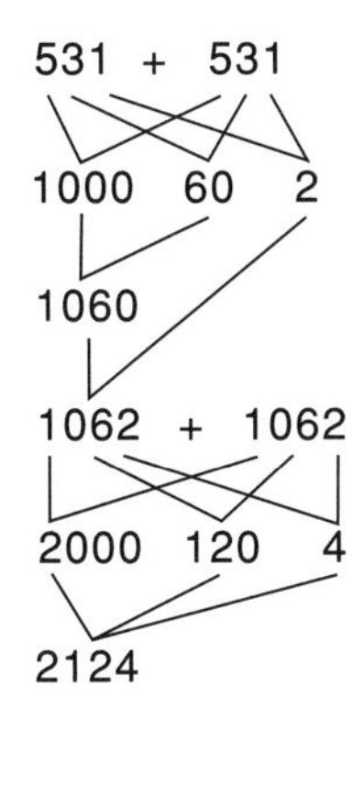

"OK. Four times 531 is 2124. Five times 531 is 2124 plus 531. It's 2655."

Fig. 19.5. Joan's strategy of partitioning a number into nondecade numbers

this strategy for the problem, "If there are 15 boxes with 177 apples in each box, how many apples are there altogether?"

In this example, Joan knew that 15 is 5×3 and used that fact to find 15×177 by solving 3×177 first. Her algorithm could be written as

$$15 \times 177 = (5 \times 3) \times 177 = 5 \times (3 \times 177).$$

To figure out three 177s, she combined the first two 177s by partitioning the numbers into 100s, 10s, and 1s, saying that 100 and 100 makes 200, 70 and 70 makes 140, and 7 and 7 is 14. She added the subtotals to get 354. Then she added 177 to 354 by breaking both numbers into 100s, 10s, and 1s again to get 531. After figuring out that 3×177 is 531, Joan computed five 531s. Joan decided to get four groups of 531 first and doubled 531 twice to get 2124. Then she added another 531 to 2124 and figured out that 5×531 was 2655.

Joan's invented algorithm for this problem seems to be very complex and includes more than one strategy. When we follow her reasoning, however, we realize that her main scheme for solving the problem 15×177 was to partition 15 into 3×5, solve 3×177, and then multiply the product by 5. All other algorithms were used to solve the subproblems that were created by the partitioning after the initial strategy was chosen. Joan used several different strategies to solve 3×177 and 5×531. She used the *complete number using repeated addition* strategy to solve 3×177 and used the *partitioning a number into non-*

decade numbers strategy to figure out 5 × 531 by partitioning 5 into 4 and 1. She also used the *complete number with doubling* strategy to figure out 4 × 531.

In this example, Joan used a multiplicative approach to change the problem 15 × 177 into 5 × (3 × 177), but used an additive approach to figure out 3 × 177 and 5 × 531. Her algorithm for the problem shows her flexible understanding of multiplication by using both additive and multiplicative strategies. Her algorithm also shows how complex an invented algorithm for a multiplication problem can be.

Some children use *partitioning a number into nondecade number* strategies with the distributive property. Kathy partitioned the multiplier 16 into 12 + 1 + 1 + 1 + 1 for the problem, "If there are 16 packs of gum with 5 pieces of gum in each pack, how many pieces of gum are there altogether?"

> *Kathy:* 12 × 5 is 60, so 13 × 5 is 65, 14 × 5 is 70, 15 × 5 is 75, and 16 × 5 is 80. It's 80.

In this example, Kathy partitioned 16 to use a multiplication fact, 12 × 5 = 60, which she already knew. By partitioning 16 into 12 and four 1s, she could avoid repeatedly adding sixteen 5s and solve the problem efficiently using her multiplication facts.

Partitioning a number into decade numbers

Even though partitioning the multiplier or multiplicand into nondecade numbers can make the multiplication process easier, it does not use groupings of ten. Using groupings of ten allows children to take advantage of the base-ten system in numbers. If they use their understanding of base ten to partition the multiplicand or multiplier, children could get the product of a number multiplied by a decade number very easily and use this algorithm for many multidigit multiplication problems. Children who partition the multiplier or multiplicand into decade numbers use the groups of ten to make the multiplication process easier. Jon's work illustrates how children use the groups of ten in their invented algorithms. Jon used the *partitioning a number into decade numbers* strategy in solving the problem, "If there are 43 floors with 61 offices on each floor, how many offices are there?" In solving the problem, Jon vertically wrote forty 61s and drew horizontal lines after every ten 61s (fig. 19.6). Even though he wrote down all 61s, Jon knew that ten 61s makes 610 and did not actually add ten 61s. He merely wrote 610 after every group of ten 61s and added the four 610s. Then he wrote three more 61s and added them. Finally Jon solved the problem by adding the sum of four 610s and the sum of three 61s. Jon's algorithm shows how he makes use of his base-ten understanding in solving this multiplication problem and how the use of the product of decade numbers makes multiplication procedures more efficient.

Partitioning both numbers into decade numbers

This invented algorithm closely parallels the conventional algorithm. Children using the *partitioning both numbers into decade numbers* strategy split both multiplier and multiplicand into decade numbers, perform each multiplication, and add the subproducts. Some children use this algorithm by calculating partial products. For example, Katie solved 26 × 39 by having four partial products. She said that 26 times 39 is like 20 × 30, 20 × 9, 6 × 30, and 6 × 9 because 20 times 39 is 20 × 30 plus 20 × 9 and 6 × 39 is 6 × 30 plus 6 × 9.

Sometimes children partition both the multiplier and multiplicand into decade numbers in very creative ways. In the following example, James partitioned both multiplicand and multiplier into decade numbers in an interesting way to solve the problem, "If there are 17 boxes of apples with 177 apples in each box, how many apples are there altogether?" In solving this problem, James partitioned the multiplicand 177 into 7, 70, and 100 and the multiplier 17 into 10 and 7 (fig. 19.7). He then multiplied 7, 70, 100 by 10 and added them up. James then partitioned 7 into 5 and 2 and solved the product of 5 × 177 by halving the product of 10 × 177. James added 1770 and the product of 5 × 177 to get 2655. Finally he added two more 177s to 2655 and got 3009.

As we have seen in Joan's algorithm, James used more than one strategy in solving this problem. His main scheme was to partition both 17 and 177 into decade numbers so that he could reduce the complexity of the problem. In solving 7 × 177, James did not partition 177 into decade

```
 61
 61
 61
 61
 61
 61
 61
 61
 61
 61    610
---
 61
 61
 61
 61
 61
 61
 61
 61
 61
 61    610
---
 61
 61
 61
 61
 61
 61
 61
 61
 61
 61    610
---
 61
 61
 61
 61
 61
 61
 61
 61
 61
 61    610
---   ----
 61   2440
 61    183
 61   ----
---   2623
183
```

Fig. 19.6. Jon's strategy of partitioning a number into decade numbers

```
177 × 17
7 × 10 = 70
70 × 10 = 700
100 × 10 = 1000
1000 + 700 + 70 = 1770

1770 ÷ 2 = 885
1770 + 885 = 2655
177 + 177 = 354
2655 + 354 = 3009
```

Fig. 19.7. James's strategy of partitioning both numbers into decade numbers

numbers again but partitioned 7 into 5 + 2. To find 5 × 177, he cleverly used the product of 10 × 177 by halving it. James's invented algorithm may not perfectly fit into the *partitioning both numbers into decade numbers* category. It shows, however, that children easily adapt their algorithms depending on the numbers in the problems and the knowledge available to them. In this example James showed his good number sense and flexibility depending on the numbers in the problem as well as his understanding of base ten and multiplication.

Compensating Strategy

Children adjust numbers in certain problems on the basis of the special characteristics of the number combinations. Sometimes they adjust both multiplier and multiplicand, and other times they adjust only one of the numbers in the problem. They then make corresponding adjustments later if necessary. When they adjust both numbers, those numbers might be doubled or halved to make the calculation easier or to use some multiplication facts that children already know. Strategies adjusting both numbers are frequently observed to be used in problems involving 5. In the following example, Isaac adjusted both multiplicand and multiplier in solving the problem, "If there are 5 bags of beads with 250 beads in each bag, how many beads are there altogether?"

Isaac: Since I am multiplying by 5, I can split 250 in half and multiply it by 10. Half of 250 is 125, and 125 times 10 is 1250. It's 1250.

Isaac knew that he would get the same answer if he halved one number and doubled the other when he multiplied the two numbers. Since multiplying a number by 10 was easy for him, Isaac decided to halve 250 and compensate by doubling 5 to 10.

When they adjust only one of two numbers, children adjust one number up or down to the nearest decade number. Then they make corresponding adjustments later on to figure out their solution. Children are frequently observed to adjust one number when they have a number close to a decade number in the problem. In the following example, Heidi used the *compensating* strategy by adjusting only one number in solving the problem, "If there are 17 bags of M&M's with 70 M&M's in each bag, how many M&M's are there altogether?"

Heidi: Seventeen 70s, mmm, I can do 20 times 70 instead. That's 1400. I need to take 210 away because I went over by three 70s. So it's 1400 take away 210. 1400, 1200, 1190. That's 1190.

(She wrote, 17 × 70 → 20 × 70 → 1400 – 210 → 1190.)

Heidi changed the problem 17 × 70 to 20 × 70 to use her understanding of base ten. She figured out 20 × 70 and made a corresponding adjustment later on by taking away three 70s.

Conclusions from the Observations

The analyses presented in this paper provide evidence that children can and do invent multiplication algorithms for multidigit multiplication problems when given the opportunity to do so. The examples in this paper illustrate a variety of invented algorithms from *direct modeling* to *partitioning numbers* and *compensating* strategies. Through the process of inventing algorithms, children develop a deeper and more flexible understanding of multiplication.

These analyses also suggest that the algorithms children invent provide us with a window into how they think about the additive or multiplicative composition of numbers in solving multiplication problems. We found that children increasingly developed their number sense as they had more opportunities to invent multiplication algorithms.

Implications for Instruction

Students in almost all the classrooms we observed invented a variety of algorithms for multiplication problems. Some children invented more efficient and abstract algorithms than others. The students showed that their understanding of multiplication was related to repeated addition by using the *complete number* strategy. By doubling or partitioning numbers, they shortened repeated addition processes. Children used their number sense and recalled known multiplication facts both in the *partitioning numbers into nondecade numbers* strategy and the *compensating* strategy. Also they showed sophisticated base-ten understanding by capitalizing on the use of multiples of 10 in the *partitioning numbers into decade numbers* strategy. Students' understanding of base ten and their number sense appeared to be closely related to their generation of invented algorithms for multidigit multiplication. A teacher's choice of numbers for multiplication problems appeared to be related to the kind of algorithm children invent. For example, multiples of 5 and numbers close to a decade number were observed to encourage children to use the *compensating* strategy.

Understanding Multiples of 10

As we have seen in the *partitioning numbers into decade numbers* strategy, the use of multiples of 10 generates very powerful multiplication algorithms. Children's invented algorithms that partition numbers into sets of ten avoid long addition procedures and allow them to think of number compositions. It is important for teachers to help students see the role that multiples of 10 play in multiplication instruction.

Number Sense

By giving students opportunities to invent algorithms, teachers can help them develop their number sense. Teachers can help children by choosing numbers that will encourage multiplying numbers by partitioning numbers into decade or nondecade numbers and putting them together. Discussions about the variety of ways of multiplying numbers give students a great opportunity to develop their number sense.

The Development of Invented Algorithms

Many children in the study developed their invented algorithms for multidigit multiplication problems in a sequence from *direct modeling* to *complete number* to *partitioning numbers into nondecade numbers* to *partitioning numbers into decade numbers.* This analysis provides a possible developmental path that teachers could use as a reference as they assist students in inventing algorithms. The descriptions of the algorithm types can help teachers identify and understand their own students' inventions. When teachers understand children's invented algorithms and their developmental paths, they can help students move to more sophisticated algorithms.

REFERENCES

Fuson, Karen C., Diana Wearne, James C. Hiebert, Hanlie G. Murray, Pieter G. Human, Alwyn I. Olivier, Thomas P. Carpenter, and Elizabeth Fennema. "Children's Conceptual Structures for Multidigit Numbers and Methods of Multidigit Addition and Subtraction." *Journal for Research in Mathematics Education* 28 (1997): 130–62.

National Council of Teachers of Mathematics. *Curriculum and Evaluation Standards for School Mathematics.* Reston, Va.: National Council of Teachers of Mathematics, 1989.

20

The "Write" Way to Mathematical Understanding

David J. Whitin

Phyllis E. Whitin

ALTHOUGH algorithms provide an efficient route to securing correct answers, children often do not understand how or why these procedures work. Merely directing children to follow the traditional rules for algorithms in a lockstep fashion is like expecting children to arrive without having made the journey (Barnes 1995). However, we have found that encouraging children to write about the process can be a valuable way for learners to make sense of these algorithms for themselves (Countryman1992). Teachers can use writing to assess their students' understanding, foster number sense, and extend mathematical conversations. The following stories from a fourth-grade classroom illustrate these important benefits.

Writing about Base-Ten Blocks to Assess Students' Understanding

Base-ten blocks are an important tool for demonstrating important mathematical concepts and properties. Danielle wrote about these blocks to show her understanding of the distributive property in the multiplication problem 72 × 25 (fig. 20.1). Her illustration and her writing help to connect the traditional algorithm to a concrete model of the process. The class had first solved several problems together. Danielle adopted the use of arrows (that we teachers had used in a class lesson) and incorporated them into her own drawing to show which sets of numbers she was multiplying together. She demonstrated an understanding of place value when she multiplied "5 longs by 7 and got 350" and "7 flats by 2 and got 1400." She described the dimensions of each of the four rectangular areas according to their appropriate lengths and connected those individual areas to the numbers in the

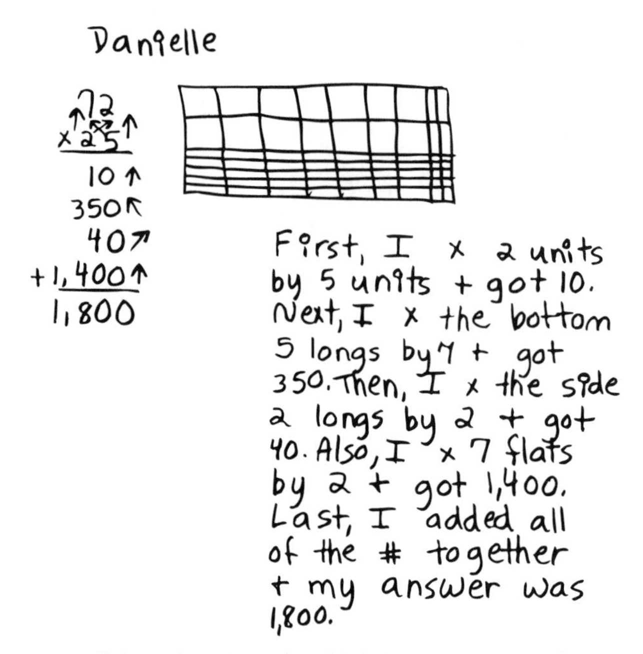

Fig. 20.1. Danielle's explanation of multiplying 72 × 25 using base-ten blocks

algorithm. Her illustration, algorithm, and written text were all important avenues for her to express her understanding of this operation and the concepts that lie behind it.

Nikki used the base-ten blocks to explain her understanding of the process of renaming and exchanging numbers. She wrote a lengthy response to 67 + 58 = 125:

> I traded my 67 and my 58 into a hundred square, and then I took the 7 and the 8 and traded them into a stick of ten. I put 6 + 5 sticks of ten and that equaled ten but I had 1 extra. How I got 125 is that I put the 67 and the 58 together to make 100 … I had 1 extra so that is 1 hundred and 1[ten]. Then when I had those two numbers 7 and 8, well, I put 2 of the little cubes to the 8 and that equals 10, so I traded that 10 for a stick of 10, so that equals 1 hundred and 2 tens. I had 5 left and that equals 125.

There are several interesting features in Nikki's explanation. First, she decided to trade in tens for a hundred as her initial step in the process, added the units next, and finally added the tens. We supported the children's efforts to solve problems in different ways, and Nikki's solution reflects her autonomy as a problem solver. Second, Nikki used her own language, such as *the stick of ten* to describe the process. Third, she added specific details to her solution by describing how she renamed 7 as 5 + 2 and added the 2 to the 8 to make 10 units.

WRITING ABOUT PERSONAL STORIES TO ASSESS STUDENTS' UNDERSTANDING

We also encouraged the students to write stories that incorporated various contexts for regrouping with tens. At first we brainstormed items that are grouped into sets of ten: packs of stickers, different kinds of candy, pencils, money, bubble gum, and so on. The children's stories reflected these various contexts. Deidre wrote about 83 – 28 = 55: "There were ten stickers in a pack and there was 8 packs and 3 extra stickers so there was 83 stickers. Well, the next day a lady came in [the store] and wanted 28 stickers. So the man gave her 2 packs and the 3 extra and that's all. 'Well,' the woman said, 'I need 5 more. I only have 23.' Well, the man opened a pack and gave her 5 more and had 55 left." Deidre's strategy is interesting because she subtracted the tens first and then subtracted the ones in a two-part process. The exchanging was portrayed as breaking open another pack of stickers. The familiar context of stickers helped to support this inventive solution strategy.

Brent devised a similar strategy when he used the context of pencils to solve the same problem, 83 – 28 = 55: "I had 83 pencils. I had to give Sam 28 but I did not have enough out of a pack. So I took three whole bags. I gave him all of them and he gave me two back." Brent also started by subtracting tens but decided to give away more than necessary (we related this strategy to paying for items in a store). He explained that 83 – 30 = 53, but the two he received back made his final answer 55. Here again, exchanging occurred when a pack of pencils had to be broken open to make the final subtraction.

Finally, Amanda used the familiar context of money to solve this same subtraction problem (see fig. 20.2): "First, you can't take 8 from 3, so 8 said,

First you can't take 8 away from 3 so 8 said I'll lend you a 10¢ (dime) witch will make that 8 into 7 dimes and add the 10¢ with the three 1¢ 1¢ 1¢ (a penny) then you count up from 8 and it will come to 5¢ (nide) then you count up from 2 to 7 and it will come to five.

Fig. 20.2. Amanda's explanation of 83 – 28

'I'll lend you a ... (dime), [which] will make that 8 into 7 dimes and add the [dime] with the three [pennies]. Then you count up from 8 and it will come to [a nickel] then you count up from 2 to 7 and it will come to five." Amanda wrote in a conversational style, employed the strategy of missing addends to do the subtraction, and used the appropriate metaphor of lending to describe the exchanging process. All these familiar contexts helped the children invent different ways to describe how the algorithm worked.

Using Writing to Develop Number Sense

Writing about algorithms can also encourage the strategic, flexible use of those algorithms by promoting the development of number sense. It is common for children to treat each place in multidigit examples as an independent problem rather than view the entire problem holistically. To address this concern, we developed the strategy of asking the students to make predictions about the problem before they worked with the blocks. This strategy proved to be particularly helpful in developing flexibility in the use of an algorithm.

For example, when Jesse was given the problem 265 – 138, he wrote, "Before I started I saw that I couldn't subtract 8 units from 5 units. So first I took away the flat. Then I took a long and gave it to the units and subtracted both of them" (see fig. 20.3). Because Jesse took the time to examine the problem and record his initial assessment of it, he decided to abandon the traditional right-to-left algorithm. He could see that the subtraction of the hundreds would not be affected by the tens and ones. By temporarily ignoring the necessary regrouping, Jesse could arrive at a rough estimate of the difference. Jesse's solution illustrates a bridge between paper-and-pencil work and mental mathematics that can be helpful when it makes sense to calculate a global answer and then fine-tune it.

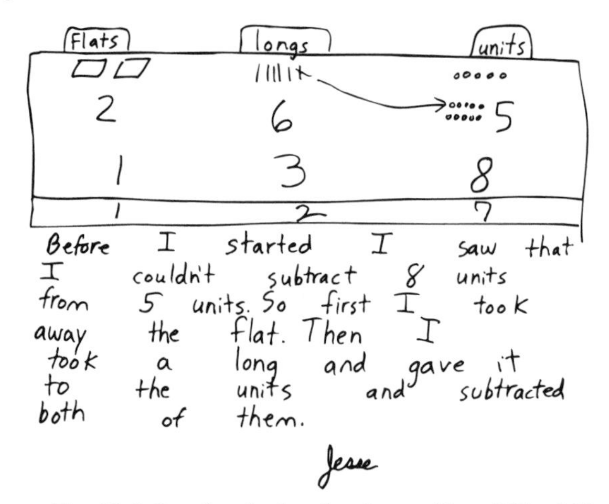

Fig. 20.3. Jesse's solution for the problem 265 – 138

Like Jesse, Jeremy viewed a problem holistically in order to make a prediction about his solution. He wrote the following about the problem 310 – 185 = 125:

> In my head I know that two of my flats are gone because it says to take one flat away, and I have one long but I need to take eight longs away. I have to take the flat away and get ten longs. Then I have to take eight longs away so I have two longs left. In my head I know that my number is going to be odd because a half of ten is five.

Jeremy shows us that he could envision the trading that would be necessary in both the hundreds and tens columns of his problem. This prediction helped him estimate a difference of about 100. The prompt, "Write about what was going through your head," was an invitation for Jeremy to share his personal interest in odd and even numbers as well.

Writing about her predictions also helped Sara evaluate her answer for reasonableness. After writing about her solution to 247 – 163, she remarked aloud, "It might be surprising at first to have such a small answer when you start with such a large number, but you could see already that you were going to use both flats." Writing about the problem gave Sara time to consider her estimate before calculating the answer. When looking at the problem as a whole, she realized that both flats would be removed. Even though the final answer might have seemed "surprising," Sara had taken the time to analyze why her answer made sense.

Colby faced a problem that was similar to Sara's. Although he did not record his predictions beforehand, he did use writing to reflect on his initial confusion. After solving 212 – 168 correctly through a diagram of blocks, he admitted aloud that he had not expected the answer to be so small. Studying his diagram, he realized that he had used one flat to trade for tens and he had removed the other to indicate subtraction. He decided that he should write about the revision he made in his thinking.

> At first I thought that my answer was going to be 100 and some units, but it wasn't because you have to take a hundred first and once you do your trading stuff you have no tens, and you have to take 6 tens from nothing, and you only have a 100 more to spare. First I took away a ten, I put ten units on my board and took away 8 from that and then this is when it gets hard. You have to take a hundred off your board and trade it for ten longs and subtract 6 from that and that is four and you get 44 all together.

Colby showed a willingness to record parts of the process that were unexpected and difficult for him. This kind of reflection gives learners the confidence to become instruments for their own learning (Whitin forthcoming). When asked why it might be helpful to keep a record of the "hard part," Colby replied, "Because next time don't get mad at that part because you'll probably get it right anyway." Writing enabled Colby to share the revision in his thinking and to build his confidence as a problem solver. Writing enabled all four children to reflect on their problems more holistically and analyze why their answer made sense.

Writing can be a valuable tool for helping teachers track the sense-making strategies of learners. Ryan solved 213 – 147 in two different ways. After he solved the problem with the base-ten blocks, he recorded a unique mental-computation strategy: "I pretend the problem is 213 – 113 = 100. 47 is 34 more than 13. 100 – 34 = 66." Since we had developed a class ritual for naming strategies to honor the inventiveness of the thinker, we asked Ryan, "What does this strategy remind you of?" He paused a few seconds and replied "Humpty Dumpty" because the breaking apart and the putting together of numbers was like the actions of "all the king's men" trying to put Humpty together again. Metaphors can be an important way for learners to express mathematical ideas (Whitin and Whitin 1996).

We then asked Ryan, "Have you always solved problems that way or is this a new strategy for you?" When he admitted that this was a new strategy, we asked him to record this changed thinking in his journal. He wrote, "I used to do the problem like it was, but now I pretend the problem is different, and sort of make it right like Humpty Dumpty and how they put him together again." We then asked him to draw a picture to show how this metaphorical description worked (see fig. 20.4). His illustration shows how he renamed the subtrahend of 147 as 113 + 34 and then subtracted those two parts separately. We were demonstrating that journals have the potential to provide a historical record of children's growth as problem solvers.

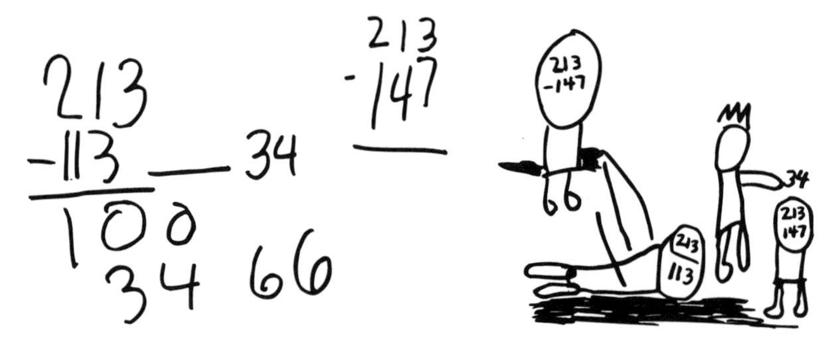

Fig. 20.4. Ryan's illustration of his "Humpty Dumpty" strategy for solving 213 – 147

USING WRITING TO PROMOTE MATHEMATICAL CONVERSATIONS

We thought it was important for students to share the thinking in their mathematics journals with other students on a regular basis. One of the strategies that we developed to address this need was to copy different journal entries on a piece of paper and distribute a copy to each member of the class (see figs. 20.5 and 20.6). We then invited the children to write a response to one of the entries. Chris was intrigued with Scott's strategy for 324 ÷ 6 (see fig. 20.5). In this problem Scott distributed the 3 hundreds to the six people of the divisor by giving each person "half of a flat." He then traded "two tens for 20 ones," which gave each person an additional four units for a total of 54. Chris wrote back to Scott, "I like what Scott did by halving the flats. I wouldn't want to use it because I don't understand it as much as I understand mine. I can see what he did. I have never used it till now. 385 ÷ 6 = 64, r 1. I did it!" Chris gave each of six people 5 tens (half a flat) and then gave each of them another ten from the set of 8. He then traded in the remaining 2 tens for 20 units and distributed the 25 units among six people. Although a bit hesitant at first, Chris was interested enough in Scott's strategy to try it out for himself. As children are exposed to a repertoire of strategies, they realize that they have a wide range of options for solving problems and become more confident about taking some risks and testing out some of these alternative strategies in their own ways.

Scott: $6\overline{)324}$ First I gave each person a half of a flat which is 5 tens. I traded two tens for 20 ones, and each person got 4 ones.

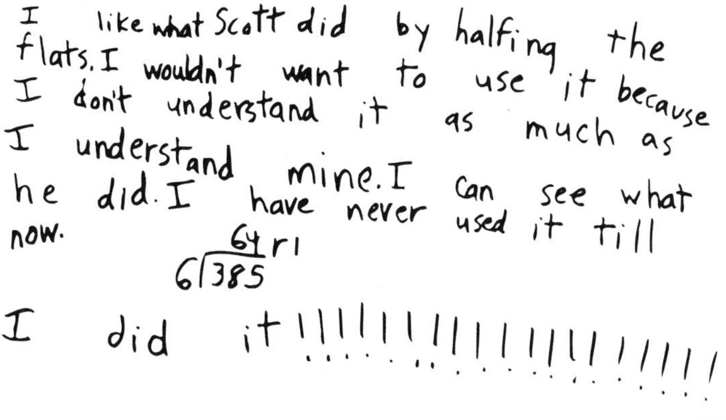

I like what Scott did by halfing the flats. I wouldn't want to use it because I don't understand it as much as I understand mine. I can see what he did. I have never used it till now.

$6\overline{)385}$ = 64 r1

I did it!!!!!!!!!!!!!!!!!!!!

Fig. 20.5. Chris's response to Scott's explanation of his solution to $6\overline{)324}$

William found Rett's strategy quite interesting (see fig. 20.6) and wrote the following response: "Rett: I like how you did the problem and how did you know by looking at the problem that you needed to trade two longs for twenty units. I've done a problem like this when my mom gave me this problem—475 ÷ 5. I looked at it and I knew that I couldn't give 5 people 1 flat." William found a common element in these problems: whenever he noticed that the hundreds digit of the dividend was one less than the digit of the divisor, he knew he was going to have to trade his flats for longs, which would give him a quotient less than 100. Sharing strategies as a classroom community enables learners to note similarities among problems and become more adept at selecting useful strategies for solving problems.

Rett: $\overset{54\text{r}0}{6\overline{)324}}$ First I put 3 flats, 2 longs, and 4 ones and I had to divide them up into 6 people. Right when I looked at the problem I knew I would have to trade for longs and then I knew that I would have to trade 2 longs for twenty ones. After that I gave everybody an equal share. Everybody got five longs and 4 ones and there was no remainder.

Rett: I like how you did the problem and how did you know by looking at the problem that you needed to trade 2 longs for twenty units. I've done a problem like this when my mom gave me this problem - $5\overline{)475}$ I looked at it and I knew that I couldn't give 5 people 1 flat.

Fig. 20.6. William's response to Rett's strategy for solving $6\overline{)324}$

A FINAL REFLECTION

Writing can be a powerful avenue for children to clarify and reflect on their own thinking. It can also serve to guide our own teaching by giving us a window into the children's reasoning and understanding. We became more knowledgeable about the children's understanding of place value and of the exchanging process and about their use of estimation strategies.

When we asked the children themselves to reflect on the potential of writing to learn mathematics, they emphasized several benefits. Lily wrote, "When I write I get lots of ideas of what else I want to say." William described writing as a way "to get your imagination going," and Jonathan wrote, "When you write about math, people know what you're talking about; you get more ideas; you get good grades; you get smart; when you speak out people know you" (see fig. 20.7). Thus, the children appreciated that writing enabled them to generate new ideas and to make their voices heard. It gave them the opportunity to forge their own paths toward mathematical understanding.

Fig. 20.7. Jonathan's reflection on the benefits of writing to learn mathematics

REFERENCES

Barnes, Douglas. "Talking and Learning in Classrooms: An Introduction." In *Primary Voices* 16 (January 1995): 2–7.

Countryman, Joan. *Writing to Learn Mathematics.* Portsmouth, N.H.: Heinemann, 1992.

Whitin, David J., and Phyllis Whitin. "Fostering Metaphorical Thinking through Children's Literature." In *Communication in Mathematics, K–12 and Beyond,* 1996 Yearbook of the National Council of Teachers of Mathematics, edited by Portia C. Elliott, pp. 60–65. Reston, Va.: National Council of Teachers of Mathematics, 1996.

Whitin, Phyllis. "Fueled by Surprise." In *Teacher's Voices: Language Arts,* edited by Judith Lindfors and Jane Townsend. Urbana, Ill.: National Council of Teachers of English, forthcoming.

Part 4

21

Letting Fraction Algorithms Emerge through Problem Solving

DeAnn Huinker

MS. VARNER asked her fifth-grade class, "If I have three-fourths of a pound of cheese in one package and seven-eighths of a pound of cheese in another package, do I have enough to serve twelve people so it is fair with everyone getting one-eighth of a pound of cheese?" Some of the students immediately pulled out their fractions strips, other students began to write or draw on paper, and a few students were visualizing the problem in their minds.

Lakesha explained her approach. "I got out my eighths strip and my fourths strip and put seven eighths together with three fourths. Then I put my whole strip over it. I knew this was the same as eight eighths and then I still had five eighths showing, so that's thirteen eighths. You can serve thirteen people." By laying the whole strip over the three fourths and some of the eighths, Lakesha saw that three-fourths plus two-eighths was the same amount as one whole.

Thomas visualized the problem. He explained, "I remembered that two eighths was the same amount as one fourth, so I just counted. Three fourths is the same as two eighths, four eighths, six eighths, and then I added the seven eighths and got thirteen eighths. You can feed thirteen people."

Nicole's written work is shown in figure 21.1. She explained, "I needed another fourth to make one whole and I know that two eighths is the same as one fourth, so I added the two eighths to the three fourths and that gave me a whole. Then I had five eighths left over. Well, one whole is the same as eight eighths and then I added the five eighths and that gave me thirteen eighths. You can give everyone an eighth of a pound and still have one-eighth pound of cheese left for yourself."

This type of episode became a common occurrence in Ms. Varner's classroom once she decided to change her approach to fraction instruction. In prior years, Ms. Varner and her colleague, Mr. Brown, taught fractions "by the book." This was the first year that both teachers put the textbook on the shelf and attempted to use a problem-solving approach for their fraction instruction. The teachers were aware of research showing that children were capable of inventing their own algorithms for whole numbers when they were encouraged to devise their own ways of solving problems (Burns 1994; Kamii, Lewis, and Livingston 1993; Madell 1985). If a similar approach was used with fractions, could children invent their own algorithms for operations with fractions? This was the question that prompted these two fifth-grade teachers and me to become an instructional team investigating a radical change in the teaching of fractions.

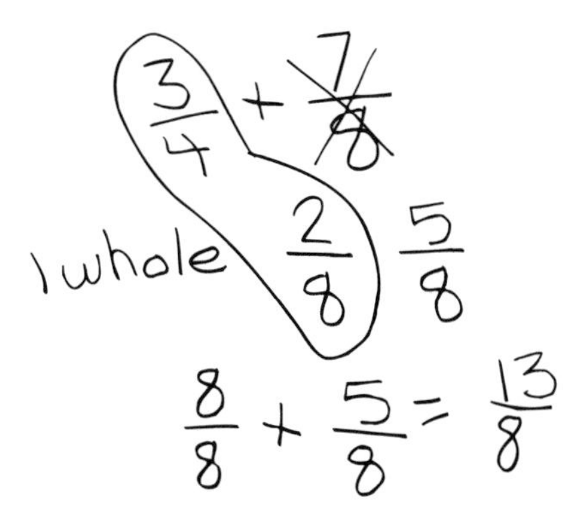

Fig. 21.1. Nicole's "make a whole" procedure

Mr. Brown and Ms. Varner teach in a large urban school system. Sixty-nine percent of the students at their school are members of minority groups, and 68 percent receive free lunch. The standardized test scores for the fifth-grade students indicated that 54 percent were performing at or above the national average. The teachers expressed their concern over not teaching the algorithms as presented in their textbooks. Would the students really be able to solve problems if they were not directly taught algorithms? Would they be prepared for sixth grade? The teachers were willing to try a new approach, but at the same time they were cautious.

Guiding Principles

"Problem solving is not a distinct topic but a process that should permeate the entire program and provide the context in which concepts and skills can be learned" (NCTM 1989, p. 23). The instructional team sought to put this statement into practice within the context of learning fraction operations. They decided not to tell their students "how to" compare, add, subtract, multiply, or divide fractions but rather to focus on developing meaning for fraction operations and to see what algorithms might emerge from students solving problems. In rethinking the teaching of fractions, the instructional team established five guiding principles:

1. *Begin instruction with problem situations.* Beginning with word problems furnishes a context for students to use concrete objects and diagrams to explore and make sense of fraction concepts and operations. At first the teacher can pose the word problems, but as soon as possible students should begin to pose and solve their own problems and those of one another.

2. *Make connections among representations.* Continual emphasis needs to be placed on connecting real-world, oral language, concrete, pictorial, and

symbolic representations. Students who are able to make these connections demonstrate lasting ability to use their mathematical knowledge flexibly to solve word problems (Lesh, Post, and Behr 1987). Students who can easily translate from one representation to another are able to use the representations as tools to approach problems from several different perspectives.

3. *Emphasize "big ideas."* This emphasis promotes coherence throughout instruction and assessment on critical aspects of fraction knowledge. The following list of big ideas was adapted and extended from an initial list presented by Payne, Towsley, and Huinker (1990).

- Wholes can be broken into equal-sized parts and reassembled into wholes. Sometimes there are not enough parts to make a complete whole. Sometimes more than a whole can be made.
- Equal-sized parts have special names. The names tell the size of each part and exactly how many equal-sized parts fit into one whole.
- A specific amount can have many names.
- The symbol (e.g., 3/4) comes from the oral language (e.g., three-fourths). The top number (numerator) tells the number of parts considered. The bottom number (denominator) tells how many equal-sized parts are in one whole.
- Adding or subtracting fractions may involve renaming or trading fractions until the parts are the same size. Sometimes only one fraction needs renaming, sometimes both need renaming, and sometimes neither needs renaming.
- Number sentences are a written record of operating on fractions.
- Fractions and whole numbers can be divided by cutting or breaking them into smaller equal-sized parts. Sometimes an amount needs to be shared equally among a specific number of groups (partitive division situations). Other times the amount needs to be separated into specific same-sized groups or packages; this may result in making complete packages or part of a package (measurement division situations).
- Multiplying fractions involves combining same-sized groups, finding part of a group, or both.

4. *Integrate the study of fraction topics.* A quick perusal of the chapters in many textbooks reveals that instruction on fractions has been fragmented into discrete skills. Traditional instruction progresses from working with common denominators to related denominators and then to unrelated denominators for fractions less than 1. Then the entire cycle is repeated for mixed numbers. Similar fragmentation occurs for multiplication and division. This fragmenting results in numerous isolated procedures for students to remember. Integrating topics, such as working interchangeably with fractions less than, equal to, and greater than 1, should occur from the first day

of instruction. When beginning the addition or subtraction of fractions, there is no need to work initially only with common denominators but rather work concurrently with common and related and unrelated denominators. This integration of fraction topics seems to occur naturally in the context of problem situations.

5. *Stress the use of "benchmarks" and mental computation throughout instruction.* Benchmarks contribute to students' number sense by allowing them to understand the relative magnitude of fractions and to develop an intuitive feel for them. Students can use these benchmarks to compare fractions by mentally reasoning about their relative size. Initial attention should be given to knowing which fractions are close to zero, equal to one whole, less than one whole, and greater than one whole. Then students should examine which fractions are equal to one-half, less than one-half, and greater than one-half.

EXPLORING OPERATIONS ON FRACTIONS

The students were not taught any algorithms during this four-week instructional unit but instead were given many opportunities to figure out their own ways to deal with fraction comparisons and operations in problem-solving situations. Students were encouraged to communicate their reasoning by sharing their strategies with one another as they worked in pairs or small groups and participated in whole-class discussions. The algorithms that emerged consisted of mental strategies as well as procedures for using manipulatives, drawing diagrams, and using symbols. The instructional activities were modified from Payne, Towsley, and Huinker (1990), Ott, Snook, and Gibson (1991), and Sweetland (1984).

Addition and Subtraction

Instruction began by developing meaning for fraction concepts integrated with developing meaning for fraction operations. The following discussion occurred on the first day of instruction in one of the fifth-grade classrooms. It illustrates how naturally fraction topics can be integrated. The students had just folded fraction strips out of paper rectangles to represent halves, fourths, and eighths.

Mr. Brown: One partner should pretend that you have three-fourths of a candy bar. Show me that with your fraction strips. Now the other partner should pretend you have one-half of a candy bar. Show me that with your fraction strips. Now, figure out who has the larger amount of candy bar and tell me how you figured that out.

Kendra: She does, because we matched them up and I could see she has more.

Mr. Brown: How much does she have?

Kendra: Three-fourths.

Mr. Brown: So three-fourths of a candy bar is more than one-half.

Other strategies were shared and discussed. Then Mr. Brown posed a new problem.

Mr. Brown: Now I want everyone to figure out how much more candy bar your partner has.

Robert: I have one-fourth more than Lakesha. When we matched them up, we could see that this piece was one-fourth of a candy bar.

Again the reasoning of several students was discussed.

Mr. Brown: Now I want you to figure out how much you have together. So, put the three-fourths candy bar together with the one-half candy bar and figure out how much that is.

This problem was more difficult for the students, until they realized it was more than one whole candy bar.

Jamese: We lined up our candy bars and then I put a whole candy bar next to it and I could see that it was one whole candy bar and another piece. And then we realized that the other piece was a fourth of a candy bar. So we have one whole candy bar and one-fourth.

The scenario above illustrates the type of interactions that occurred regularly in the classrooms. In general, word problems were posed each day for students to investigate in pairs. Next the students shared their reasoning with the entire class. Then the teacher would focus on making connections among oral language, concrete, and real-world representations. During the first few days of the unit, fractions were written according to the oral language—for example, "3 fourths + 1 half." This use of written oral language helped to prevent the tendency of students to just add the numerators and the denominators. The use of fraction symbols was deliberately delayed until students demonstrated some understanding of fraction quantities. Fraction symbols were then introduced as a shortcut method for writing fractions.

Two algorithms for adding fractions emerged. The "make a whole" algorithm used by Jamese, Lakesha, and Nicole presented earlier (fig. 21.1) was very common among the students. This algorithm involved determining whether the two fractions combined made more than one whole. If so, then the students added until they could make one whole and then figured out how much was remaining. Initially, the students performed this algorithm with the fraction strips by making comparisons, but eventually students began to transfer this thinking to work with symbols, as Nicole did in figure 21.1.

"Renaming" or "trading" for same-sized pieces was the other procedure that emerged. The emphasis on the big idea—that an amount can have many different names—formed the basis for this procedure. Students were often challenged to find other names for fractions like one-half, three-fourths, or one-third with their fraction kits. These kits consisted of fraction strips for halves, fourths, eighths, thirds, sixths, ninths, twelfths, fifths, and tenths. The students used a trial-and-error approach as they matched their strips to find equivalent fractions. When adding fractions, some of the students used this matching technique to help them determine which fractions could be traded in order to get same-sized pieces. For example, when adding two-thirds and one-sixth, Robert determined he could trade his two-thirds strip for four sixths.

Eventually students began to transfer this "renaming" approach to work with symbols, as shown in D'Juan's written work in figure 21.2. The technique of writing the number sentence in a horizontal format and then crossing out and renaming fractions was very natural for the students. This raised a question for the instruction team: Should students be asked to use the vertical orientation found in most textbooks for adding and subtracting fractions? We decided to ask the students. After showing the students how number sentences with fractions could be written in a vertical orientation, it was unanimous that this was much more confusing and that the students preferred the horizontal format.

The approaches for subtraction were more varied. The first algorithm to emerge was "matching" with the fraction strips. For example, if John had two-thirds of a pound of cheese and then he used one-half pound to make a pizza, how much cheese would be remaining? To figure out the difference, students would compare the fraction strips and then examine the extra piece. Sometimes it was easy to see the value of the extra piece, but sometimes it was not obvious. Then the students would use a trial-and-error approach comparing the extra piece to the other fraction strips until they found a match. In this example, the extra piece would eventually be matched to one-sixth. This procedure was used mainly with the paper strips or with diagrams but did not transfer easily to work with symbols.

The students also used a "renaming" algorithm for subtracting fractions. With this procedure, the students would rename one or both fractions until they had the same-sized pieces and then they could subtract. This approach was similar to the standard algorithm, but as

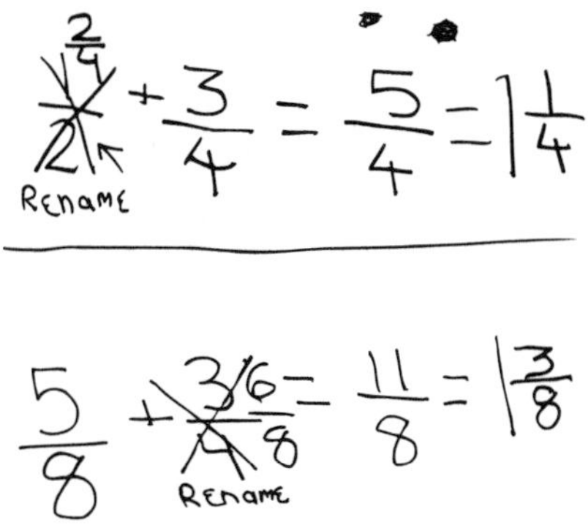

Fig. 21.2. D'Juan's "renaming" procedure

with addition, the students preferred using a horizontal format and the crossing-out method when using symbols.

Two algorithms also emerged for subtraction with mixed numbers. During one lesson, students were asked to work on this problem and to record their thinking in writing: "Pretend you had a new bag of bird seed that weighs two and one-third pounds and then you filled up the empty bird feeder, which holds five-sixths of a pound. How much bird seed would you still have?" Christopher's thinking is shown in figure 21.3. He used a "subtract and subtract" procedure. First he subtracted what he could from the fraction part of the mixed number, and then he subtracted the rest from the remaining whole number. Marie thought about the problem differently. She used a "subtract and then add" procedure. This procedure was more common among the students. She explained, "I subtracted the five-sixths from the two whole pounds, which is one and one-sixth pounds. Then I added that to the one-third. I know that one-third is the same as two-sixths, so that equals one and three-sixths."

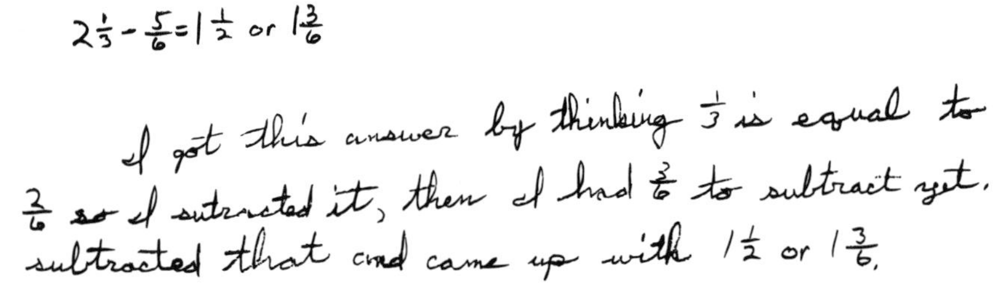

$2\frac{1}{3} - \frac{5}{6} = 1\frac{1}{2}$ or $1\frac{3}{6}$

I got this answer by thinking $\frac{1}{3}$ is equal to $\frac{2}{6}$ so I sutracted it, then I had $\frac{3}{6}$ to subtract yet. subtracted that and came up with $1\frac{1}{2}$ or $1\frac{3}{6}$.

Fig. 21.3. Christopher's thinking

Multiplication and Division

A division problem was also integrated into the first lesson of the unit. The students had just finished adding one-half and three-fourths when the teacher posed this problem for them: "Right now you and your partner have one whole candy bar and one-fourth of a candy bar. I want you to share that amount between the two of you and see if you can tell me what part of a whole candy bar each person gets." It was fairly easy for the students to share the paper candy by tearing it apart and giving each person three pieces (see fig. 21.4). Sherrie offered the following observation: She explained that each person got one-fourth piece of a candy bar, one-half of the one-half piece, and one-half of the other one-fourth piece. Now the difficulty arose in deciding how much of a candy bar that was all together. By comparing their fraction strips, the students eventually realized that one-half was the same amount as two-fourths and that one-fourth was the same amount as two-eighths. After some more exploration, the students finally arrived at an answer of five-eighths of a candy bar for each person.

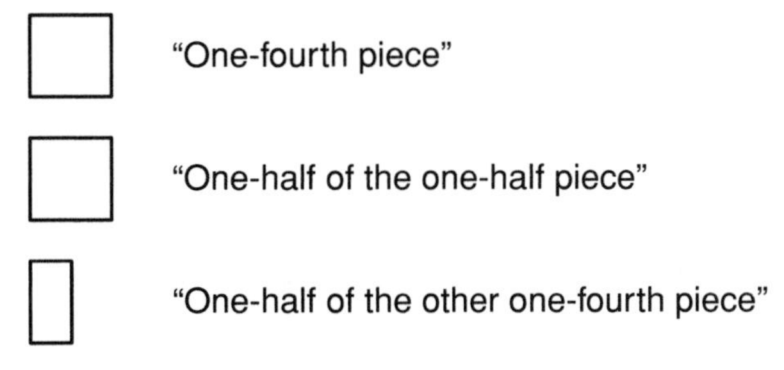

Fig. 21.4. Sherrie's three pieces

During the third and fourth weeks of instruction, word problems continued to be posed each day with a focus on multiplication and division situations. Students first explored measurement division situations, then partitive division situations, and finally multiplication situations. Students used their fraction strips to model the situations directly and then shared their approaches and reasoning with one another. After a problem was solved and discussed, the mathematical operation was identified and symbols were used to write a number sentence to record the relationships among the numbers in the problem. Many students seemed to shift fairly quickly from the use of fraction strips to drawing diagrams. This seemed to occur because the diagrams allowed students to indicate groupings easily, as shown in figures 21.5–21.8. The algorithms that emerged for multiplication and division generally consisted of procedures for using diagrams or mental reasoning.

The students used a "grouping" procedure for measurement division situations. They would first tear the fraction strips or mark their diagrams into the appropriate-sized pieces and then group them. Felisha's work is shown in figure 21.5. She realized everything needed to be renamed to fourths, and then she circled groups of three fourths. Some of the measurement division problems posed to the students had remainders. For example, if this problem had stated that Jim had four candy bars, then he would have been able to give three-fourths of a candy bar to five of his friends and would have had one-fourth of a candy bar left over for himself. At the end of this lesson, Felisha wrote in her journal, "I learned that you can divide fractions today. The ones we had were easy. You have to put the fractions into groups like you do when you divide just numbers."

Two procedures were used for partitive division situations. For example, if you have one and one-half candy bars to share equally among four people, how much would each person get? The first procedure that emerged involved "cutting and sharing." This is the method Helen used in figure 21.6. She cut the whole candy bar into four equal pieces and then she cut the one-half into four equal pieces. After further prompting and exploration, Helen arrived at an answer of three-eighths. The major difficulty with this procedure involved figuring out how much of a whole candy bar each person received after the sharing had been completed.

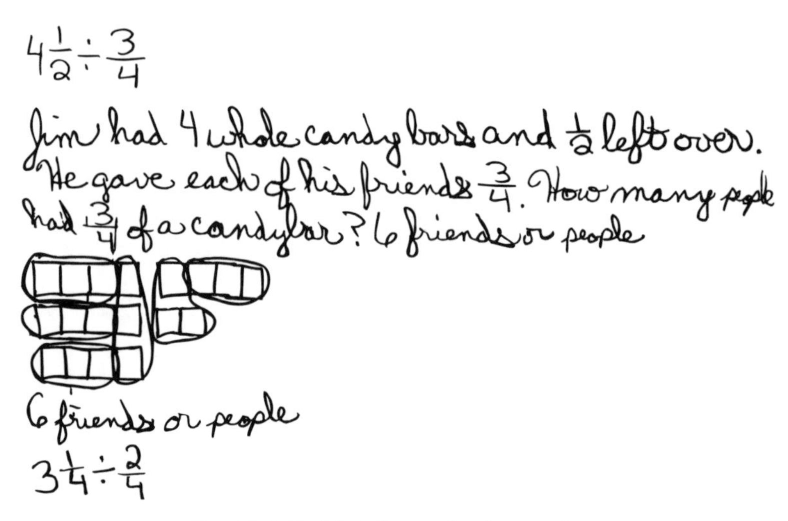

Fig. 21.5. Felisha's "grouping" procedure

Fig. 21.6. Helen's "cutting and sharing" procedure

The other procedure that emerged was "check and rename." Daniel's work is shown in figure 21.7. He explained his reasoning as follows: "I tried halves, but that didn't work. Only three people could each get a half. Then I tried thirds, but that didn't work. Then I tried sixths. Each person could get two sixths, but I still had some left over. Then I tried eighths, and that worked because I made twelve eighths and gave each person three of them." Some students carried out this procedure with their fraction strips, whereas others were able to reason mentally.

Two procedures for multiplication situations emerged. Many students used a "repeated addition" procedure in which they would model the problem with

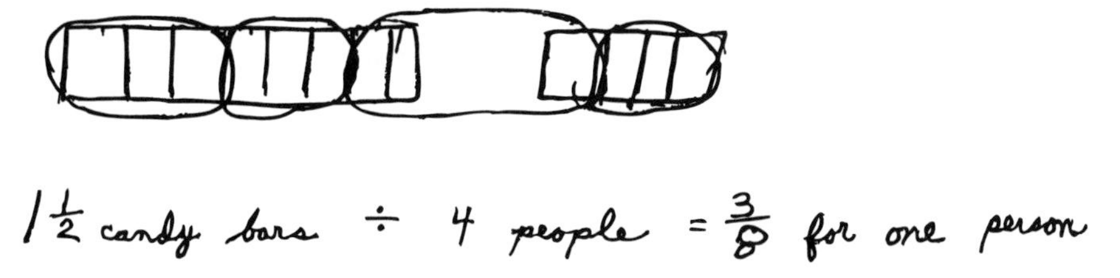

Fig. 21.7. Daniel's "check and rename" procedure

fraction strips or diagrams and then add or multiply to find the total number of pieces. The other approach involved grouping the pieces to "make wholes" as shown in Selena's work in figure 21.8. The instructional team decided not to pursue situations that involved multiplying a fraction by a fraction. The students had been working hard for four weeks and had shown tremendous growth in their understanding of fraction operations. The team decided it was time to reflect on the overall progress of the students.

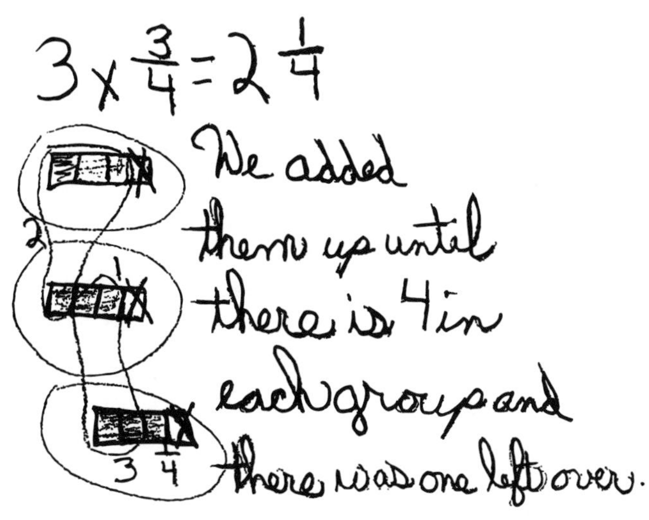

Fig. 21.8. Selena's "make wholes" procedure

ASSESSMENT OF STUDENTS' UNDERSTANDING

Students' knowledge of fractions was assessed prior to and following instruction. All forty-seven students completed a written test, and several students from each class were individually interviewed. Students demonstrated very limited understandings of fraction concepts and fraction operations prior to instruction. The assessment results revealed that all students made some progress in their understanding of fractions. The following are selected results from the assessments that illustrate some of the changes.

On the pretest, only 10 percent of the students could accurately explain why 4/3 was greater than 11/12. The reasoning used by 77 percent of the students involved stating that 11/12 was greater "because it had larger numbers." On the posttest, 66 percent of the students accurately explained why 4/3 was greater, with most stating that four-thirds was more than one whole and eleven-twelfths was less than one whole, showing their use of the "benchmark" of 1.

Before this unit of instruction, most students simply added or subtracted the numerators and denominators. Only 9 percent of the students correctly solved 3/8 + 7/8 on the pretest compared to 83 percent on the posttest. On the pretest, 2 percent of the students correctly solved 5/6 – 1/3 compared to 53 percent on the posttest.

Students were also asked to solve some word problems on the written test. Only 6 percent correctly solved a word problem representing 3 × 5/12 on the pretest compared to 49 percent on the posttest. For the word problem representing 3 ÷ 3/4, 6 percent responded correctly on the pretest compared to 55 percent on the posttest.

The interviews clearly showed that most students initially performed manipulations on fraction symbols that reflected their knowledge of whole numbers. By the end of instruction, the students dealt with the fraction symbols as representing some quantity and did not merely manipulate the symbols without meaning. The teachers would have liked the achievement of their students to have been higher, but they were extremely pleased and surprised with the reasoning ability of their students. As Mr. Brown and Ms. Varner thought about their instructional practices in previous years as compared to this inquiry into teaching, they acknowledged that their previous students did not really understand, or enjoy learning about, fractions. The emphasis had always been on memorizing many different procedures. With the problem-solving and conceptual emphasis they experienced during the past four weeks, these students felt a firm foundation had been laid on which they could continue building during the remainder of the school year.

ADVANTAGES OF STUDENT-INVENTED ALGORITHMS FOR FRACTIONS

If students are to truly believe that mathematics, and fractions in particular, makes sense, instruction must allow students to invent their own ways to operate on fractions rather than memorizing and practicing the procedures imposed by the teacher or textbook. In fact, premature focus on algorithms may actually be harmful to children, since it distorts their beliefs about the nature of mathematics—that mathematics is primarily memorizing rather than reasoning (Burns 1994; Kamii and Lewis 1993; Madell 1985). Marilyn Burns (1994) strongly stated her opinion about the risk of teaching standard algorithms:

> Imposing the standard arithmetic algorithms on children is pedagogically risky. It interferes with their learning, and it can give students the idea that mathematics is a collection of mysterious and often magical rules and procedures that must be memorized and practiced. Teaching children sequences of prescribed steps for computing focuses their attention on following the steps, rather than on making sense of numerical situations. (P. 472)

Several advantages of allowing students to invent their own algorithms were demonstrated during this inquiry into teaching fractions to two classes of fifth-grade students. The major advantages were as follows:

- Students developed an interest in solving and posing word problems with fractions.
- Students became flexible in their choice of strategy for solving fraction word problems and computation exercises.
- Students became more proficient in translating among real-world, concrete, pictorial, oral language, and symbolic representations.
- Students became accustomed to communicating and justifying their thinking and reasoning.

These students constructed intuitive quantitative understandings of fraction concepts and operations in the context of solving and posing realistic problems. Students can be successful and confident with fractions when they are given opportunities to focus on the big ideas of fractions in a manner that integrates, rather than fragments, fraction topics and that emphasizes connections among real-world experiences, concrete models and diagrams, oral language, and symbols. Many students wrote in their journals, "I learned fractions can be fun" or "I learned that fractions are easy," but Shannon summed it up best by stating, "I learned that you can do mostly anything with fractions."

REFERENCES

Burns, Marilyn. "Arithmetic: The Last Holdout." *Phi Delta Kappan* 75 (February 1994): 471–76.

Kamii, Constance, and Barbara A. Lewis. "The Harmful Effects of Algorithms in Primary Arithmetic." *Teaching K–8* 23 (January 1993): 36–38.

Kamii, Constance, Barbara A. Lewis, and Sally Jones Livingston. "Primary Arithmetic: Children Inventing Their Own Procedures." *Arithmetic Teacher* 41 (December 1993): 200–203.

Lesh, Richard, Tom Post, and Merlyn Behr. "Representations and Translations among Representations in Mathematics Learning and Problem Solving." In *Problems of Representation in the Teaching and Learning of Mathematics,* edited by Claude Janvier, pp. 33–40. Hillsdale, N.J.: Lawrence Erlbaum Associates, 1987.

Madell, Rob. "Children's Natural Process." *Arithmetic Teacher* 32 (March 1985): 20–22.

National Council of Teachers of Mathematics. *Curriculum and Evaluation Standards for School Mathematics.* Reston, Va.: National Council of Teachers of Mathematics, 1989.

Ott, Jack M., Daniel L. Snook, and Diana L. Gibson. "Understanding Partitive Division of Fractions." *Arithmetic Teacher 39* (October 1991): 7–11.

Payne, Joseph N., Ann E. Towsley, and DeAnn M. Huinker. "Fractions and Decimals." In *Mathematics for the Young Child,* edited by Joseph N. Payne, pp. 175–200. Reston, Va.: National Council of Teachers of Mathematics, 1990.

Sweetland, Robert. "Understanding Multiplication of Fractions." *Arithmetic Teacher* 32 (September 1984): 48–52.

22

Developing Algorithms for Adding and Subtracting Fractions

Glenda Lappan

Mary K. Bouck

KNOWING how to combine quantities and remove a quantity is a mathematical skill that is basic to understanding the world around us. These operations of addition and subtraction are a part of the mathematical experiences of students throughout grades K–12. As students proceed through the grades, however, they encounter different types of numbers. This paper focuses on the middle grades and the development of an understanding of adding and subtracting fractions. The contexts and the ideas for the approach were developed by the Connected Mathematics Project (CMP) team. This team started work in 1991 to produce for all students in grades 6–8 a connected, coherent middle school mathematics curriculum that is balanced across areas of mathematics.

One question developers of materials such as CMP's ask is, How can curriculum materials help students develop their understanding and ability to add and subtract with different kinds of numbers, fractions in particular? In the past, students learned to add and subtract fractions by memorizing the algorithms presented in their textbook. These algorithms were taught over several lessons, with each lesson adding a new dimension. The difficulty with this approach is that some students never learn the algorithms and others who can carry out an algorithm never understand why they perform certain steps and develop little sense of when the algorithm is useful to solve a problem. The intent of good mathematics instruction is not just that students can add and subtract fractions by using efficient algorithms but that

The Connected Mathematics Project was funded by the National Science Foundation. However, the philosophy and opinions expressed in the materials are not necessarily those of the foundation. The principal investigators on the project were Glenda Lappan, James T. Fey, William M. Fitzgerald, Susan N. Friel, and Elizabeth Phillips.

students understand what it means to add and subtract fractions and can assess what is a reasonable solution for a given problem. In order to foster such understanding and skill, teachers need to engage students in designing algorithms and analyzing the way various invented algorithms work and how they relate the meaning of the operation to the numbers involved.

How can curriculum materials support this agenda? What might a series of problems look like that would help students develop their own algorithms to compute with fractions? How can we support this development so that students understand not only how to get a solution but why they do what they do and why their solution is reasonable?

The following problems from the CMP unit entitled *Bits and Pieces II* (Lappan et al. 1996a) offer a set of engaging situations whose solutions require adding and subtracting fractions. The contexts of the problems help students make sense of how to put fractions together and take them apart. The problems come from a unit that follows an intensive focus in *Bits and Pieces I* on developing the meaning of fractions, decimals, and percents and flexibility in moving from one form of representation to another. *Bits and Pieces II* does not teach a particular algorithm for computing. Strategies for computing with fractions come from class discussions after the students have worked on the problems. It is true that letting the students wrestle with making sense of situations takes more time than showing them an algorithm, but the payoff in the long run is that students learn to think and to reason about mathematical situations. The invented algorithms of students are often very efficient and with a teacher's help can become powerful, generalizable methods. In work such as this, the teacher does not leave to chance the students' developing of algorithms. He or she is actively engaged in asking questions, giving needed information, and pointing the students toward important considerations.

In this set of problems, the responsibility for developing algorithms rests not only with the curriculum materials but also with the teacher. The teacher guides the classroom interactions and through these interactions judges when the students are ready to look for an efficient algorithm for adding and subtracting.

The mathematical and problem-solving goals for this set of problems are the following (Lappan et al. 1996a, 1996b):

- To develop strategies for adding and subtracting fractions
- To understand when addition or subtraction is the appropriate operation to help make sense of a situation
- To develop ways of modeling sums and differences
- To use estimation to make sense of adding or subtracting fractions
- To reinforce the understanding of equivalence of fractions
- To look for patterns and to generalize the patterns observed

PROBLEM 1: MAKING THE MAP

This problem engages students with an area model for fractions. The students have to name what fraction of a section of land each person owns. The problem uses both fractions of a section of land and measures of the land in acres. This allows questions that call for multiplying a whole number by a fraction.

When Tupelo Township was founded, the land area was divided into sections that could be farmed. Each section is 1 square mile of land, that is, 1 mile long on each edge. There are 640 acres of land in a 1-square-mile section. Here is a diagram of two adjacent sections of land, sections 18 and 19. The name of the owner is on each part of the section.

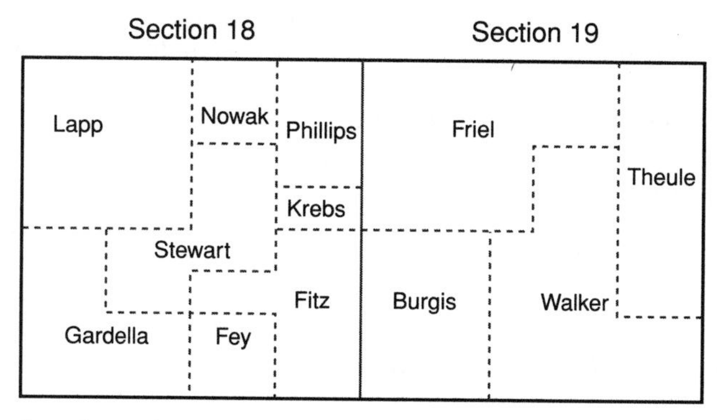

A. What fraction of a section does each person own? Explain your reasoning.

B. How many acres of land does each person own? Explain your reasoning.

This problem does not ask students to do any formal computation but instead to review their experiences of naming fractional parts of a whole. Yet many students do solve this problem by using their intuitive understanding of combining and removing amounts. The problem allows students to take what they know and build on that understanding.

Here is a sample of the kinds of questions a teacher can use to involve students in the problem:

- How many sections of land are being considered in this problem?
- Who owns the largest part of a section? Does anyone own a whole section?
- What are some of the things you have to think about before naming the fractional part of a section of land that each person owns?

Given a copy of the land-ownership map, students often use a ruler to divide the sections into pieces of equal size so they can find the fractional amounts of the land that each person owns. A common strategy that students use is to divide each section into sixty-fourths (see fig. 22.1). When the students were asked how they decided that each section had to be divided into sixty-fourths, Heidi said the following:

> I didn't know it had to be sixty-fourths when I started. What I did was take the property lines and I extended them until they reached the section borders. I did this for each property line. Then, I looked to see where else I needed to add lines to make each piece in the section the same size, and I added those lines. When I got all done, the pieces were all the same size, and then I could just count how many of the pieces each person had out of the sixty-four pieces that made a whole section.

When asked how she kept the property lines clear, Heidi said that she drew in the new lines with a different color so that it was possible to tell where the original property lines were.

Another strategy students use is to think about how the pieces of land relate to a whole section and then how they relate to one another. Students see that section 18 can be divided into four large parcels. Since Lapp has one of those parcels, Lapp has 1/4 of a section. Gardella has one of those four parcels (1/4) less a quarter of it. The missing quarter of a quarter section is 1/16 of a section, so Gardella has 3/16 of a section because 1/4 – 1/16 is 3/16. Fey has a plot equivalent to the amount that Gardella is missing, so Fey has 1/16 of a section. Nowak has a plot of the same size as Fey's, so Nowak has 1/16 of a section. Krebs owns a portion that is half the size of Nowak's plot, so Krebs has half of 1/16 of a section, or 1/32 of a section. Phillips has a plot that is the size of Nowak's and Krebs's together, so Phillips must have 3/32 of a section. The students continue with this type of reasoning, combining and removing parts of a section, until all the landowners' properties are labeled with fractional names.

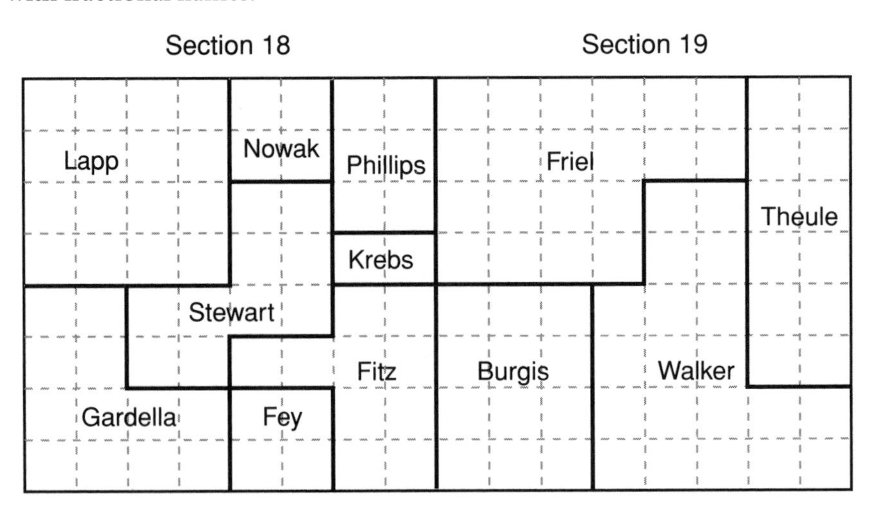

Fig. 22.1. A common strategy for problem 1 is to divide each section into sixty-fourths.

Other students name all plots in terms of Krebs's portion. Then they have to figure out what fraction of a section a "Krebs" is to make the final statements about each person's land. For example, Stewart owns five "Krebs" of land. Since a "Krebs" is 1/32 of a section, Stewart owns 5/32 of a section.

Even though the students are not saying it formally, notice that they are finding common denominators by dividing areas and adding and subtracting only the numerators of fractions. Also notice that they are doing so not because they are being told but rather because it makes sense to them.

PROBLEM 2: REMAKING THE MAP

This problem involves students in sorting out a set of buying and selling transactions. The information given about the transactions serves as clues to help figure out what plots of land are sold to whom. Several options need to be considered as students work through the problem. The growing list of constraints leads to a reduction in the number of possibilities. At the same time, the problem presents a situation that requires reasoning to solve and engages students in computation with fractions by both combining and removing fractional parts.

Some of the owners sell land to other owners of land in sections 18 and 19. What sales could have taken place for these results to happen?

- When the sales are all completed, four persons—Theule, Fey, Phillips, and Gardella—own all the land in the two sections.
- Theule bought from one person and now owns land equivalent to 1/2 of one section of land.
- Fey bought from three people and now owns the equivalent of 13/32 of one section of land.
- Gardella now owns the equivalent of 1/2 of a section of land.
- Phillips now owns all the rest of the land in the two sections.
- Each of the four owners can walk around all of his or her land without having to cross onto another person's land.

A. What transactions took place? Explain exactly which tracts of land Theule, Fey, Phillips, and Gardella bought and how you know that you are correct.

B. To show your thinking, draw a new map of the two sections outlining the land belonging to each of the four owners. In each section show a fraction sentence that explains how each section fits the clues given. Tell how many acres each person now owns.

The final solution to this problem must take into account all the constraints given. For example, the second constraint is that Theule bought from one person and now owns land equivalent to 1/2 of one section of land. If we consider only this clue, she could have bought from Friel or

Walker. As we read through the remainder of the clues, we must eliminate one of these solutions.

Working in groups of three or four students is a good arrangement for the size and complexity of this problem. The context provides a lot of practice with adding simple fractions, even though students may not recognize how much practice is embedded.

Let's consider the information one piece at a time to find a solution:

- We know that all the owners except Theule, Fey, Phillips, and Gardella have sold all their land. If Theule now owns a plot that is equivalent to 1/2 of a section, then she must have bought 5/16 of a section to add to her original plot that is 3/16 of a section. Theule could have bought either the Friel or the Walker plot.
- Fey bought from three people, so he could have bought from any person who has sold land. Fey started out with 1/16 and needed to add to get a total of 13/32. Fey could have bought 5/32 from Stewart, 5/32 from Fitz, and 1/32 from Krebs or 8/32 from Lapp, 2/32 from Nowak, and 1/32 from Krebs.
- Gardella has a total of 1/2 of a section, so she must have bought 5/16 of a section. She could have purchased Friel's plot, Walker's plot, both Lapp's and Nowak's plots, or both Stewart's and Fitz's plots.
- Phillips owns all the rest of the land in the two sections. If Theule and Gardella own 1/2 section each and if Fey has 13/32 of a section, then Phillips owns 19/32 of a section. Phillips has 3/32 of a section, so he must therefore purchase a total of 1/2 a section. Phillips must have bought Burgis's plot and either Friel's portion, Walker's portion, or both Lapp's and Nowak's portions or both Stewart's and Fitz's portions.
- No owner has to cross another owner's land to get to any part of his or her own. This means that Fey must have purchased Fitz's portion and two other adjacent portions that total 6/32 of a section, which must be Krebs's and Stewart's plots. Gardella must have purchased Lapp's plot and Nowak's adjacent plot. Phillips must have purchased Friel's plot along with Burgis's plot. Theule must have purchased the remaining portion, Walker's plot.

To summarize (see fig. 22.2):

1. Theule bought Walker's land: $\frac{3}{16}+\frac{5}{16}=\frac{8}{16}=\frac{1}{2}$
2. Fey bought Stewart's plus Kreb's plus Fitz's land: $\frac{2}{32}+\frac{5}{32}+\frac{1}{32}+\frac{5}{32}=\frac{13}{32}$
3. Gardella bought Lapp's plus Nowak's land: $\frac{3}{16}+\frac{4}{16}+\frac{1}{16}=\frac{8}{16}=\frac{1}{2}$
4. Phillips bought Friel's plus Burgis's land: $\frac{3}{32}+\frac{10}{32}+\frac{6}{32}=\frac{19}{32}$

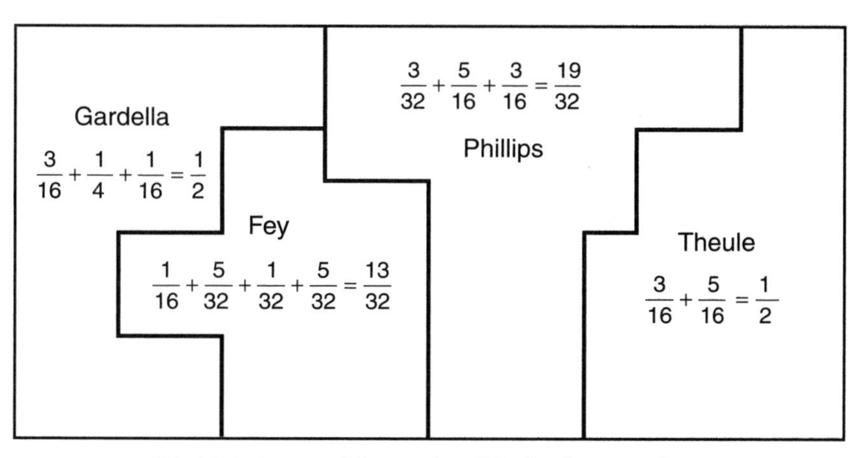

Fig. 22.2. A map of the results of the land transactions

A Conversation in a Sixth-Grade Classroom Late in the School Year

The following is part of a dialogue that took place in one classroom during the summary of this problem. The students in this school are accustomed to working with a problem-solving mathematics curriculum that asks them to work in groups and make sense of the situation without being told how to do the problems. These students have also studied equivalent fractions and naming and comparing fractions in a manner similar to the way ideas are presented here.

Teacher: I need one of the groups to start the conversation about problem 2 by explaining what transactions they think took place and why they think they are correct.

Anne: We had more than one possible answer for some of the transactions until we got through all the regulations and then we ended up with the following (*the text in fig. 22.3 is written on a large sheet of paper and displayed at the front of the room by other members of Anne's group*).

Theule — bought Walker's land $\left(\frac{3}{16}+\frac{5}{16}=\frac{8}{16}=\frac{1}{2}\right)$

Fey — bought Stewart's + Krebs's + Fitz's land $\left(\frac{2}{32}+\frac{5}{32}+\frac{1}{32}+\frac{5}{32}=\frac{13}{32}\right)$

Gardella — bought Lapp's + Nowak's land $\left(\frac{3}{16}+\frac{4}{16}+\frac{1}{16}=\frac{8}{16}=\frac{1}{2}\right)$

Phillips — bought Friel's + Burgis's land $\left(\frac{3}{32}+\frac{10}{32}+\frac{6}{32}=\frac{19}{32}\right)$

Fig. 22.3

We think we are right because we have accounted for all the land. We have shown that each person has the amount they are supposed to have. For Theule and Gardella, this is half a section each and for Fey, 13/32 of a section. We checked our work by thinking about how Theule and Gardella each own half a section, and a half plus a half is a whole section. And if you add Fey's and Phillips's land amounts together, you have 13/32 + 19/32 and that equals 32/32, which is a whole section. So, our group has accounted for all the land in the two sections.

Here is our map that shows how the land is now owned (*displays the new map*). You can see that each owner can walk around all of the land they own and not cross onto another person's property.

Teacher: Where did you get the fractions that you have written beside each person's land purchases?

Anne: We got those from our answers to problem 1. On our sheets we had that Theule owns 12/64 of a section, but that is the same as the fraction 3/16. And for Walker, she did own 20/64, but that is the same as 5/16. So that is where we got those fractions.

Teacher: How do you know that 12/64 is the same as 3/16 and that 20/64 is the same as 5/16?

Anne: Because in the other unit we did on fractions, we learned how if you multiply or divide the numerator and the denominator by the same number, you have equivalent fractions. For these fractions we could divide the numerators and denominators by 4 and get equivalent fractions.

Teacher: Why does that make sense?

Anne: Remember, we made more parts on fraction strips to look for equivalent fractions. When you multiply the denominator by a number, you are making more equal parts, but they are smaller, so you have to multiply the numerator by the same number to see how many parts will keep the amount the same.

Teacher: Why did you rename the fractions as sixteenths and not leave them as sixty-fourths?

Anne: We had done it on our map and we thought it was easier to work with fractions with denominators of sixteenths rather than sixty-fourths.

Teacher: How do you know that 3/16 and 5/16 gives you 8/16, which you say is 1/2?

Anne: Well, the denominators tell us the size of the pieces, and because the denominators are the same, then we know that for these two fractions, the size of the pieces is the same. But we have different amounts of pieces, which we know because the numerators are different and it is the numerators that tell us how many pieces we have. So we thought that if we had three pieces that were each a sixteenth and five pieces that were each a sixteenth, then we would have a total of eight pieces, each a sixteenth, and 8/16 is the same as saying you have 1/2.

Teacher: If I look at the other purchases, it seems that you have given fractional names for each person's property and those names always have the same denominators. Why did you do that?

Anne: So that we could add the amounts. We need the pieces to be the same size so that we can tell how many there are all together.

The conversation continued with the teacher's asking the class to think about what this group has presented. She asks questions like the following:

- What do others think about this group's strategy?
- Does it seem reasonable?
- Does anyone have a different answer? (If yes, explain why you think your answer is reasonable and correct for the constraints of the problem.)

PROBLEM 3: PUTTING TOGETHER PARTS

This problem has a fanciful setting that involves students in finding the amount of pizza that the pizza pirate has eaten over a given number of days. This problem is rich in mathematical ideas. Because the context is engaging and informative, students have a better chance of persevering in a fairly demanding mathematical situation. This problem not only requires adding, multiplying, and subtracting fractions but also lends itself to many different models to represent the ideas. Encountering the concept of the infinite process in situations such as this is a mathematical bonus for the student's growing maturity in mathematical thinking.

Courtney's class made a gigantic square pizza for the class party the day after the final exam. They made it a week before the party so that they could have time to study for exams. To keep the pizza fresh, they put it in the school freezer. Unfortunately, a notorious pizza pirate was lurking in the area. On the first night the pizza pirate came in and ate half the pizza. On the second night he ate half of what was left. Each night he came in and continued to eat half of what was left. He was getting very tired of pizza!

When the class went to get their pizza, what fraction of the pizza was left? Make a table or chart showing—

A. how much pizza the pirate ate each day;
B. how much total pizza he has eaten by the end of each day; and
C. how much pizza was left at the end of each day.

Write a summary of how your group found a solution to the problem. Draw any diagrams that are helpful in showing your thinking.

This is an appropriate problem to be done in groups of three or four students. In composing the written summary, groups should include in their reports all strategies that members of the group used to make sense of the computations needed to solve the problem. Looking for patterns in the amounts found for the first several days can suggest to the students a way to compute solutions for large numbers of days and, ideally, to generalize the solution to any number of days.

This problem is so rich in mathematical possibilities that teachers should take sufficient time to discuss the solutions the students find as well as the patterns that can be discovered from their tables and charts. A group's work might be organized as shown in table 22.1. Some groups did drawings to help them make sense of the problem. The drawing in figure 22.4 was given by one group of students in addition to their table.

TABLE 22.1
An Example of a Table Students Developed for the Pizza Pirate Problem

Day	Amount Eaten	Total Amount Eaten	Amount Left
1	$\frac{1}{2}$ of $1 = \frac{1}{2}$	$\frac{1}{2}$	$1 - \frac{1}{2} = \frac{1}{2}$
2	$\frac{1}{2}$ of $\frac{1}{2} = \frac{1}{4}$	$\frac{3}{4}$	$\frac{1}{2} - \frac{1}{4} = \frac{1}{4}$
3	$\frac{1}{2}$ of $\frac{1}{4} = \frac{1}{8}$	$\frac{7}{8}$	$\frac{1}{4} - \frac{1}{8} = \frac{1}{8}$
4	$\frac{1}{2}$ of $\frac{1}{8} = \frac{1}{16}$	$\frac{15}{16}$	$\frac{1}{8} - \frac{1}{16} = \frac{1}{16}$
5	$\frac{1}{2}$ of $\frac{1}{16} = \frac{1}{32}$	$\frac{31}{32}$	$\frac{1}{16} - \frac{1}{32} = \frac{1}{32}$
6	$\frac{1}{2}$ of $\frac{1}{32} = \frac{1}{64}$	$\frac{63}{64}$	$\frac{1}{32} - \frac{1}{64} = \frac{1}{64}$
7	$\frac{1}{2}$ of $\frac{1}{64} = \frac{1}{128}$	$\frac{127}{128}$	$\frac{1}{64} - \frac{1}{128} = \frac{1}{128}$

The visual representation helps many groups find the solutions for the first few days. The patterns in the table are helpful in predicting the amounts for large numbers of days. Some groups notice that the amount for each day is a sequence of unit fractions, with each denominator double the one before. When this idea comes up, teachers ask students to give arguments to show why the pattern makes sense. This occasion is an opportunity to focus

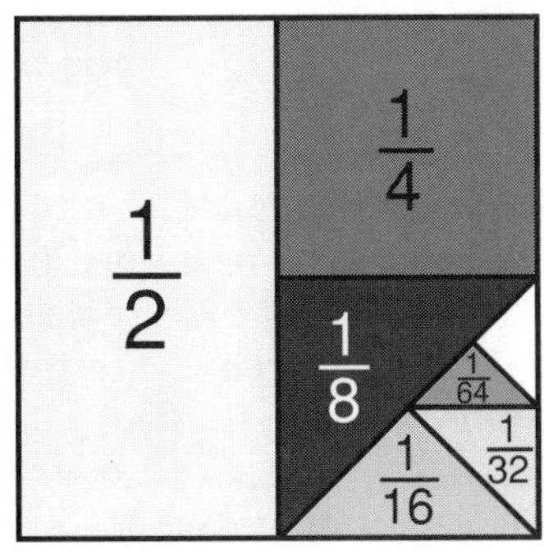

Fig. 22.4

the students' attention on what the denominator means: the number of pieces of equal size into which the whole (or the unit) has been divided. Therefore, when the pizza pirate takes half of what is left, he is really dividing the remaining piece into two equal parts. If the original pizza had been divided into parts the size of these new pieces, there would have been twice as many parts and the new pieces would have been half as large.

Students notice that the denominators are powers of 2 and that in the "Total Amount Eaten" column, the fraction representing the sum has a denominator 1 greater than the numerator. This observation makes sense because the pizza pirate always leaves one piece of the new-sized pieces on the plate, which means that the amount he has eaten so far is all but one of the new-sized pieces.

PROBLEM 4: DESIGNING AN ALGORITHM FOR ADDING AND SUBTRACTING FRACTIONS

One of the dilemmas teachers face in problem-centered teaching is deciding when and how to encourage students to make their algorithms for doing useful computations more explicit and generalizable. In our earlier work with the first three problems, we did not make an opportunity to discuss more explicitly algorithms for adding and subtracting fractions. Not doing so leaves to chance students' putting together their experiences to form an efficient algorithm. We designed the following problem to give students working in groups time to organize their collective strategies into working algorithms.

This problem asks students to reflect on and record their strategies for solving problems that require them to add or subtract fractions. Each student in a group is expected to have at least one algorithm that she or he can use for adding or subtracting. Through this experience students and the teacher can talk about what a mathematical algorithm is. Recording what students say and revising the written algorithm until the class agrees that what is written makes sense to someone reading these directions help students to understand what it means to design an algorithm and what the standards are for writing a complete description.

It is not uncommon for students to describe the standard algorithm of writing the fractions so they have common denominators and then adding or subtracting the numerators. This is a very effective means of working with fractions. Some students may have strategies that are not exactly the standard algorithm but are also effective.

The problem was presented as follows:

> In order to become skillful at handling situations that call for addition and subtraction of fractions, you need to have a good plan for carrying out the computation. In mathematics, such a plan is called an *algorithm.* For an algorithm to be useful, it has to be carefully described so that anyone who carries out the steps of your algorithm correctly will be able to add or subtract fractions and understand why each step in your plan makes sense. This problem asks you to work with your group to develop an algorithm for adding fractions and an algorithm for subtracting fractions. You may develop more than one algorithm in your group. What is important is that each member of your group understand and feel comfortable with at least one algorithm that will help add or subtract fractions. You might want to look back over the first three problems and discuss how each person in your group thought about those problems. Look for ideas that you think will help develop a plan for adding or subtracting fractions that will always work, even with mixed numbers.
>
> A. When you have an algorithm, test it on these problems:
>
> 1. $\frac{5}{8} + \frac{7}{8} =$ 2. $\frac{3}{4} - \frac{1}{8} =$
>
> 3. $\frac{5}{6} - \frac{1}{4} =$ 4. $\frac{3}{5} + \frac{5}{3} =$
>
> 5. $5\frac{4}{6} - 2\frac{1}{3} =$ 6. $3\frac{3}{4} + 7\frac{2}{9} =$
>
> B. After you have tested your algorithm for adding and subtracting fractions, revise your work if you need to.

This problem works well if students are given time to talk about their strategies first and then each person in the group writes his or her own algorithm. When all the students have finished, they share their drafts with their group and others in the group make suggestions about how each written algorithm can be improved. To have only one person in the group write will not help all students develop their reasoning and communication skills in mathematics.

Having each group share their algorithms with the class allows a variety of approaches to be discussed and allows comparisons and evaluation that often help students make more sense of adding and subtracting fractions. Some teachers make a cumulative list of the different algorithms that are given.

Sample Algorithms

Below are first drafts of algorithms, each designed by one of three students in the same class. The teacher suggested that the students write their algorithms by thinking about writing a letter to a friend. Giving students an audience for their writing is often helpful. One of the students, Kim, wrote *two* letters, one for each of her algorithms.

Jay's Algorithm

Dear Friends,

When you are finding the sum or difference of two fractions, what do you do? This is what I do. If the denominators aren't the same you should make them the same. The easiest way to do this is to think of multiples that have the denominators as factors. Let's say this is one of your denominators—8—and this is the other—5. The only multiple that I can think of that has 8 and 5 as factors is 40. So you multiply 8 by 5 to get 40 and 5 by 8 to get 40. Now you have to change the numerators too. The first numerator is 4 and the second is 2. You have to multiply it by the same number you multiplied the denominators by. $4 \times 5 = 20$ and $2 \times 8 = 16$. Now you put your fractions together and now you can find the some or difference.

$$\frac{20}{40} + \frac{16}{40} = \frac{36}{40} \qquad \frac{20}{40} - \frac{16}{40} = \frac{4}{40}$$

from,

Jay

Missy's Algorithm

Dear Jamie,

My strategy for finding the sum of two fractions is changing the numbers to fractions that have the same denominators then you add or subtract.

$$\frac{2}{3} + \frac{3}{6} = \frac{7}{6} \text{ or } 1\frac{1}{6} \qquad \frac{4}{6} + \frac{3}{6} = \frac{7}{6} \text{ or } 1\frac{1}{6}$$

Your Friend,

Missy

P.S. Another way is to use bars. Example: $\frac{3}{6} + \frac{2}{3} =$

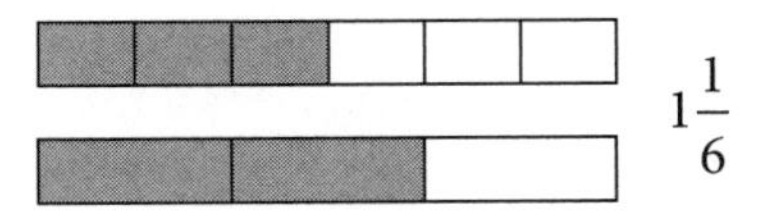

$1\frac{1}{6}$

Kim's Algorithms

Dear a friend,

To find the sum of 2 fractions first if there is a whole number(s) add them. Then if the denominators are different find a common multiple or factor by multiplying or dividing the denominator. When you have a common denominator all you have to do is add the two numerators and "carry" the denominator.

Have fun adding,
Kim

P.S. ex. $4\frac{3}{4}+1\frac{1}{8}=5\frac{7}{8}$ $\frac{3}{4}\times\frac{2}{2}=\frac{6}{8}$ $\frac{6}{8}\times\frac{1}{8}=\frac{7}{8}$

Dear a friend,

To subtract two fractions first if you have whole number(s) forget them right now. You might need to borrow for the fractions. If the denominator are different find a common multiple or factor by dividing or multiplying the. When you have a common denominator all you do is subtract the numerators and "carry" the denominators. Hut, if the second fraction is bigger you have to barrow 1 from the whole number. After you subtract the fractions subtract the whole numbers.

Have fun subtracting,
Kim

P.S. ex.

$4\frac{3}{4}-2\frac{1}{8}=2\frac{5}{8}$

$\frac{3}{4}\times\frac{2}{2}=\frac{6}{8}$ $\frac{6}{6}-\frac{1}{8}=\frac{5}{8}$

All three letters present algorithms that are close to the standard algorithm found in traditional textbooks, but the language the students use to explain their algorithms is not the traditional language. All three students talk about needing the "same" denominators in order to add or subtract, but in two of the papers, the students note that the common denominator is a common multiple of both the denominators of the fractions they are trying to add. Although all three papers need revising to improve the students' explanations and to complete the algorithms, it is apparent that all three of these students are making considerable progress in understanding adding and subtracting fractions.

SUMMARY

Kieren (1988) recommends that instruction should build on students' understanding of fractions and use objects or contexts that have students acting on something or making sense of something instead of just manipulating symbols according to a given algorithm. We believe the set of problems presented in this paper meets these criteria. His recommendation also raises new questions, such as, How much do students need to work with an algorithm before they take ownership of it? How much do students need to work with an algorithm before they can use it automatically on demand?

Although the problems above offer teachers a set of experiences for their students that allow and encourage invented algorithms, they do not shed light on the two questions we raise. Further research in classrooms where students invent algorithms and use them over time is needed to address these concerns.

As we continue to move forward in ways that seem consistent with helping students develop a deeper understanding of mathematics, all teachers, regardless of the level at which we teach, must take seriously our responsibility to document both what is powerful for students and what new and old unresolved dilemmas surface in our work. In doing so, we recognize that teaching is problem solving, that teaching continues to be and will always be a process of managing competing and conflicting demands. There is no "getting it right" in teaching, but there is the continuing process of "getting it better."

REFERENCES

Kieren, Thomas E. "Personal Knowledge of Rational Numbers: Its Intuitive and Formal Development." In *Number Concepts and Operations in the Middle Grades,* Research Agenda for Mathematics Education series, vol. 2, edited by James Hiebert and Merlyn Behr, pp.162–81. Reston, Va: National Council of Teachers of Mathematics, 1988. (Hardcover version available from Hillsdale, N.J.: Lawrence Erlbaum Associates, 1988)

Lappan, Glenda, James Fey, William Fitzgerald, Susan Friel, and Elizabeth Phillips. *Bits and Pieces II: Using Rational Numbers.* Palo Alto, Calif.: Dale Seymour Publishing Co., 1996a.

———. *Teachers Guide, Bits and Pieces II: Using Rational Numbers.* Palo Alto, Calif.: Dale Seymour Publishing Co., 1996b.

23

A Constructed Algorithm for the Division of Fractions

Janet Sharp

In order to complete arithmetic computations efficiently throughout the elementary and junior high school years, students are taught several algorithms. One traditional algorithm is the division of two common fractions. Because this "invert and multiply" algorithm does not develop naturally from using manipulatives (Borko et al. 1992), it is unlikely that children will invent their own "invert and multiply" algorithm. The purpose of this paper is to discuss a strategy to enable students to construct an alternative algorithm.

Before a student can be expected to invent this algorithm, hands-on knowledge of (1) whole-number division and (2) basic fraction concepts is essential. Division through fair sharing involves the allocation of the given amount across a set number of groups, whereas division through repeated subtraction requires successive removals of a set amount from the given amount. When using the fair-sharing approach, students divide objects among a certain number of groups. This approach is most intuitive when the number of groups is a whole number. Imagining the division of an amount across two and a half groups is perplexing to most learners. Repeated subtraction affords no such pitfalls, so this method will be applied to the discovery of the algorithm. Both pattern blocks and fraction circles provide excellent models to support students' creation of the necessary prerequisite fractional knowledge, such as equivalent fractions.

Common Denominators

The first common-denominator exercises to be explored by the student are those exercises with whole-number solutions. For example, $3/4 \div 1/4$ and $8/12 \div 2/12$ might be studied. The specific division situations to be explored will depend on the teacher's selection of hands-on materials. In the first example, three-fourths is the given amount and one-fourth is the amount that needs to be repeatedly subtracted. (Recall that fair sharing would require students' visualizing the rather daunting image of sharing three fourths across one-fourth of a set.) In this example, illustrated in figure 23.1, it is clear that the part to be repeatedly subtracted is removed a total of three times. So $3/4 \div 1/4 = 3$. Similarly, eight-twelfths divided by two-twelfths will result in four subtractions. That is, $8/12 \div 2/12 = 4$. The teacher might write these exercises, together with the solutions, with large numbers and place them about the room. The children will then have a continuous visual reference to the symbolic representations of the exercises.

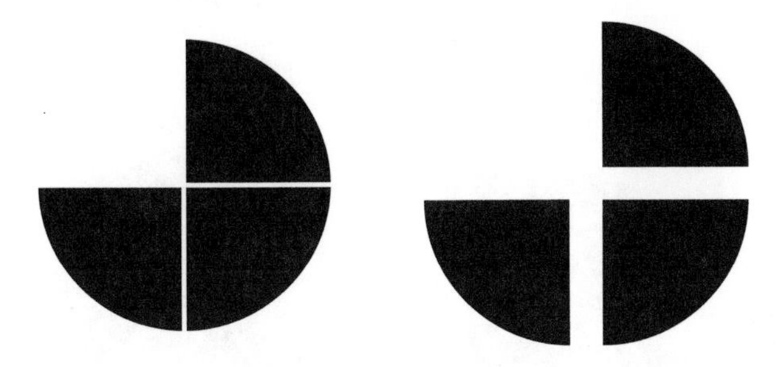

Fig. 23.1. A representation of $\frac{3}{4} \div \frac{1}{4}$

The second common-denominator exercises to be explored by the student are those exercises with nonintegral solutions greater than 1. (Exercises with solutions less than 1 are often the most difficult for children to visualize.) Examples might include $5/6 \div 2/6$ and $8/12 \div 5/12$. As shown in figure 23.2, it is clear that two sets of two sixths can easily be subtracted from the given amount of five-sixths. But after this subtraction has been completed, there is something remaining in the given amount. The students must resolve what should be done with that remainder. With whole numbers, we sometimes allow students to call this amount *the remainder* and move to the next exercise. But in this instance, we should prod the students toward a fractional settlement of the remainder. To do so, the students must recognize that the amount in the remainder is one-half of the amount *they wished to subtract.* Such a realization can often be accomplished through questioning such as "What about this amount? We *wish* we could subtract two sixths, but we have only one sixth here. What does that mean? How can we name this remainder?" Students can use manipulatives to discover a final resolution for the exercise. From five-sixths, the subtraction of two complete sets of two sixths and the subtraction of one-half of a set of two sixths can be completed. So $5/6 \div 2/6 = 2\ 1/2$. Similarly, eight-twelfths divided by five-twelfths will result in the subtraction of one complete set of five twelfths, and the remaining amount is three-fifths of a complete set of five twelfths (see fig. 23.3). Then $8/12 \div 5/12 = 1\ 3/5$. This idea may come more easily to students who have previously discovered that remainders from long division with whole numbers can be written as fractions. Once again, the teacher should display these symbolic representations of the work.

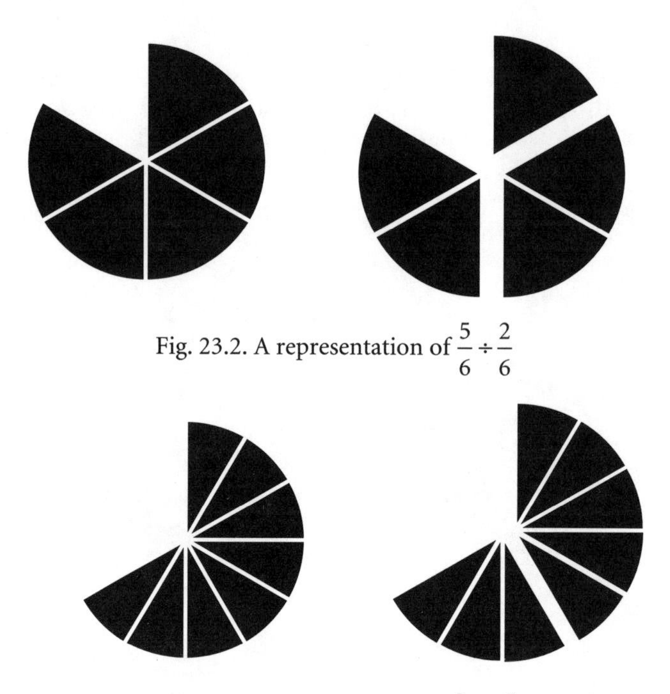

Fig. 23.2. A representation of $\frac{5}{6} \div \frac{2}{6}$

Fig. 23.3. A representation of $\frac{8}{12} \div \frac{5}{12}$

Using calculators with displays that accommodate common fractions allows children to spend their *thinking* energy making connections between whole-number division and fractional division.

UNLIKE DENOMINATORS

In order to consider an exercise requiring the division of two fractions expressed with unlike denominators, students must have an understanding of equivalent fractions. As with the preceding discussion, exercises with whole-number solutions should be explored, followed closely by exercises with non-whole-number solutions. For instance, 2/3 ÷ 1/6, 3/4 ÷ 1/6, and 1 5/6 ÷ 2/3 might be presented. To consider two-thirds divided by one-sixth, the children need to reflect on what it will mean to subtract one sixth repeatedly from two-thirds. Looking at the manipulatives, they quickly realize that there are no sixths represented in the given amount available for the necessary subtraction. However, if the students have had foundation-building experiences with equivalent fractions using manipulatives, it is typical for them to suggest that two thirds might be exchanged for four sixths. Four-sixths divided by one-sixth is then similar to the previously completed exercises. One sixth can be subtracted a total of four times. So 2/3 ÷ 1/6 is equivalent to 4/6 ÷ 1/6, and the solution is 4 (see fig. 23.4).

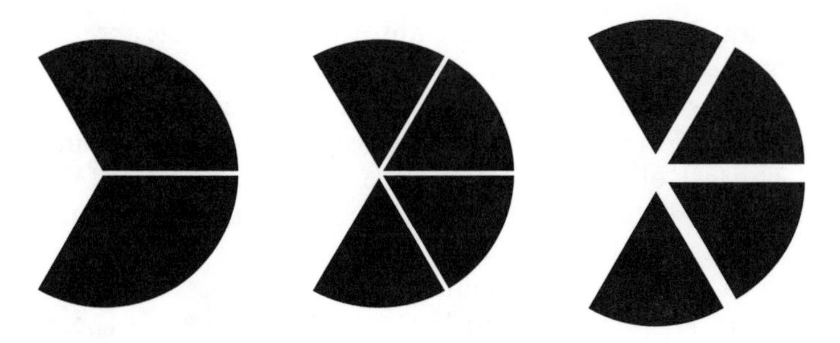

Fig. 23.4. A representation of $\frac{2}{3} \div \frac{1}{6}$

Similarly, three-fourths divided by one-sixth cannot immediately be completed. But by rewriting both fractional amounts as twelfths, the given exercise results in a familiar-looking exercise very much like those with which students have already achieved success (see fig. 23.5). From nine-twelfths, a set of two twelfths can be subtracted a total of four and one-half times. So 3/4 ÷ 1/6 is equivalent to 9/12 ÷ 2/12, and the solution is 4 1/2.

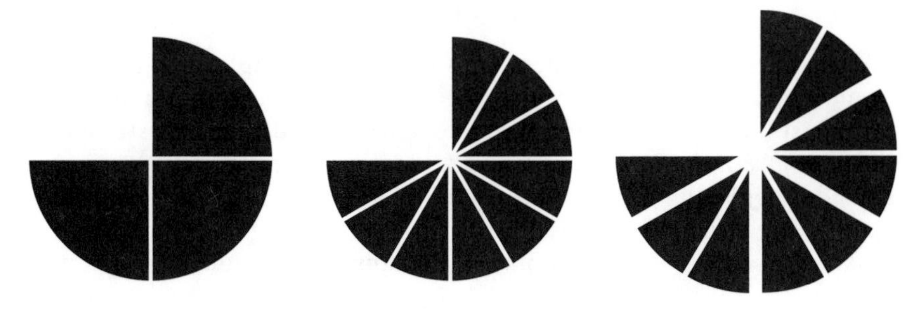

Fig. 23.5. A representation of $\frac{3}{4} \div \frac{1}{6}$

The final exercise, 1 5/6 ÷ 2/3, can be introduced by a worthwhile discussion about whether to show the one whole as six sixths together with the given five sixths. Regardless of the conclusion, the students will eventually decide to change all amounts to sixths because two-thirds, the amount to be subtracted, will not obviously be represented in the model. Once the exercise reads 11/6 ÷ 4/6, it is familiar, and the students quickly find the solution, 2 3/4. As with the previous examples, the teacher should display these results in the classroom so the symbolic representations of the exercises and solutions are visible to all the students.

SOLUTIONS LESS THAN 1

Finally, the very difficult exercises with solutions less than 1 should be considered. For example, 1/6 ÷ 3/6 might be investigated. In this example, the given amount is far less than the amount to be subtracted. In some instances, students might say that this exercise cannot be completed because the amount they wish to remove is more than the amount they are given. Such a rich situation can produce a discussion about subtraction as well as division. The task of solving the dilemma of removing a set of three sixths from a set of one sixth is not trivial (see fig. 23.6). The teacher may wish to remind students that the given amount can be viewed exactly as the remainder was viewed in an earlier example: "I would like to remove three sixths, but I have only one sixth with which to begin. What should I do?" The children should come to see that they can remove only a portion of the requested amount. That is, only one-third of the amount to be removed exists in the original set! Thus 1/6 ÷ 3/6 = 1/3.

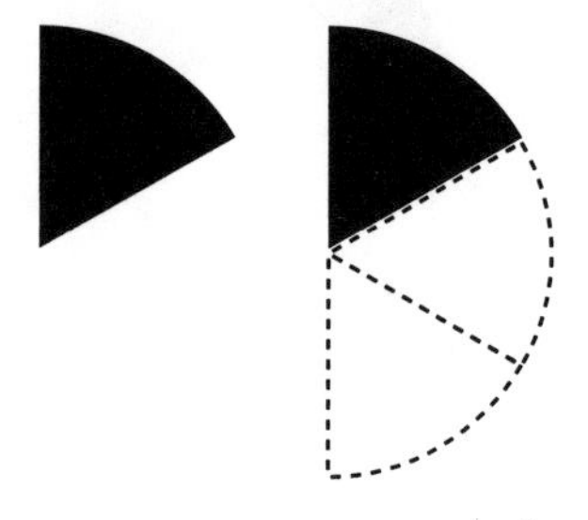

Fig. 23.6. A representation of $\frac{1}{6} \div \frac{3}{6}$

INVENTING THE ALGORITHM

After several hands-on experiences have been completed, the children should be given time to discover the algorithm that emerges from these exercises. Displaying the exercises on the classroom walls enables students to look for patterns in them (see fig. 23.7). Previous experiences with recognizing and generalizing patterns are helpful to the learners as they consider this pattern. The algorithm we wish the students to discover is, Rewrite the fractions with common denominators and then divide the resulting numerators.

$$\frac{3}{4} \div \frac{1}{4} = 3 \qquad \frac{8}{12} \div \frac{5}{12} = 1\frac{3}{5} \qquad \frac{5}{6} \div \frac{2}{6} = 2\frac{1}{2}$$

$$\frac{3}{4} \div \frac{1}{6} = \qquad \frac{2}{3} \div \frac{1}{6} =$$

$$\frac{9}{12} \div \frac{2}{12} = 4\frac{1}{2} \qquad \frac{4}{6} \div \frac{1}{6} = 4 \qquad \frac{1}{6} \div \frac{3}{6} = \frac{1}{3}$$

Fig. 23.7

Generally, at least one student will see the pattern and describe the algorithm. After all the students hear the explanation and see the pattern, a final emphasis can be made by asking them to describe the algorithm in their own words in their journals. Thereafter, we refer to the algorithm as "Kelly's recipe" or "Terry's steps," named after the first student to express the algorithm. This personalizes the procedures and sends the message that a high value is placed on the students' individual abilities to invent and justify their own algorithms. Many students who know both algorithms prefer the common-denominator algorithm because it seems to match the algorithms for the addition and subtraction of fractions, whereas the invert-and-multiply algorithm seems enigmatic.

Because it followed from whole-number operations and fraction concepts, the learning of this alternative algorithm began with comfortable and familiar activities. Using such a development for the learning of a new algorithm provided a basis not only for understanding algorithms but also for developing algebraic thinking. Much of algebra can be thought of as generalized procedures from arithmetic.

REFERENCE

Borko, Hilda, Margaret Eisenhart, Catherine A. Brown, Robert G. Underhill, Doug Jones, and Patricia C. Agard. "Learning to Teach Hard Mathematics: Do Novice Teachers and Their Instructors Give Up Too Easily?" *Journal for Research in Mathematics Education* 23 (May 1992): 194–222.

ADDITIONAL READING

Kamii, Constance."Constructivism and Beginning Arithmetic (K–2)." In *Teaching and Learning Mathematics in the 1990s,* 1990 Yearbook of the National Council of Teachers of Mathematics, edited by Thomas J. Cooney, pp. 22–30. Reston, Va.: National Council of Teachers of Mathematics, 1990.

Van de Walle, John A. *Elementary School Mathematics.* White Plains, N.Y.: Longman, 1994.

24

Dividing Fractions by Using the Ratio Table

Jonathan L. Brendefur

Ruth C. Pitingoro

Mrs. Pitingoro, a seventh-grade mathematics teacher, wanted her students to learn a more flexible way to divide fractions than the traditional "switch, flip, and multiply" algorithm. She decided to use the unit "Cereal Numbers" from the reformed curriculum *Mathematics in Context: Cereal Numbers* (1997) because it introduces students to the ratio table as a way to divide fractions. The ratio table is a representational tool that allows students to organize their work and to solve problems that deal with equivalent ratios (see Middleton and van den Heuvel-Panhuizen [1995]). But it can also be seen as an algorithm in that students can use the ratio table to generate equivalent ratios. Because the ratio table is also a representational tool, it permits students flexibility. They can manipulate the equivalent ratios by adding, subtracting, multiplying, and dividing until they have solved the problem. This flexibility is powerful for two reasons: First, the students must constantly think about what they are solving in order to know when to stop manipulating the numbers. Second, by generating equivalent ratios, the students leave behind a visible trail of their thinking. This trail helps teachers understand how their students are thinking about the problem.

Dividing Fractions in Context

This paper focuses on how students in Mrs. Pitingoro's class learned to use the ratio table to divide fractions. The students worked through the section of a unit entitled "Serving Size." The section begins with a scenario that involves students in working with Viviana, a research analyst for a cereal company, to determine the amount needed of an individual test portion for each of five different cereals and the total amount of each cereal needed for sixty testers. The test portion of each cereal is one-fourth the serving size of the cereal. The students are given the following information: (*a*) each cereal box contains 16 cups of cereal and (*b*) the serving sizes for the cereals are 1 1/2 cups for Corn Crunch 1, 1 1/4 cups for Corn Crunch 2, 1 cup for Corn Crunch 3, 3/4 cup for Corn Crunch 4, and 2/3 cup for Corn Crunch 5.

The students are first asked to find the amount for a test portion (in cups) of Corn Crunch 1. In addition, the students must determine how many boxes of Corn Crunch 1 to order for the sixty testers. The students were asked to solve this first problem by using any method they wanted. As Mrs. Pitingoro walked around the room, she noticed that her students solved the problem by using the traditional algorithm, a drawing, a calculator, or a combination of operations.

Mrs. Pitingoro had the students discuss how they understood and solved the problem. This discussion enabled the students to observe and discuss the similarities and differences among the different strategies.

LEARNING TO DIVIDE FRACTIONS BY USING THE RATIO TABLE

Next, Mrs. Pitingoro's class read how Viviana used the ratio table (see fig. 24.1) to find the amount of cereal in one test portion of Corn Crunch 2. In the story Viviana discovered that 1 1/4 cups divided among 4 people is the same as 5 cups divided among 16 people. In other words, she needed 5/16 cup for a test portion. The students were then asked to explain how Viviana's strategy worked to yield the proper amount for the test portion.

The students wrote explanations themselves and then shared them during a class discussion. The students reasoned that Viviana had multiplied both the 1 1/4 cups of cereal and the 4 people by 4 to get the 5 cups of cereal for 16 people. They then concluded that Viviana needed to get the amount of cereal needed for 1 person, so she divided the 16 people by 16. Similarly, the 5 cups of cereal needed to be divided by 16, which, the class stated, left 5/16 of a cup for every 1 person.

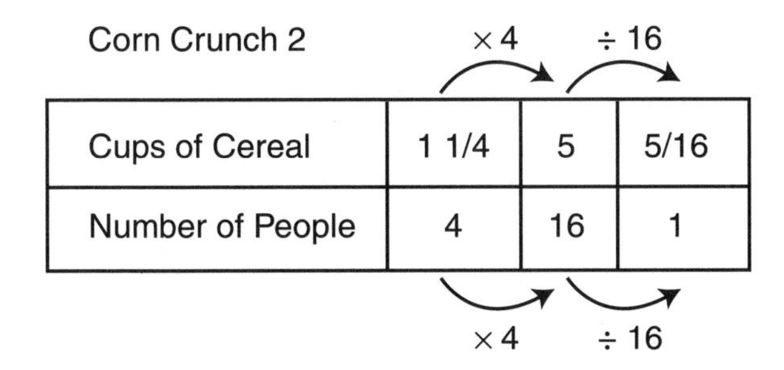

Cups of Cereal	1 1/4	5	5/16
Number of People	4	16	1

Fig. 24.1. Viviana's strategy for finding the amount of a test portion for Corn Crunch 2

After Mrs. Pitingoro was satisfied with the students' responses, she instructed them to work the next problem, which required them to use the ratio table to find (*a*) the number of cups of cereal needed for each of cereals 2 through 4, (*b*) the number of boxes of each cereal needed, and (*c*) the amount of a test portion for cereals 3 through 5. These problems enabled the students to become familiar with using the ratio table to divide fractions. The students set up the ratio table similarly to the way Viviana had set up her ratio table, but each student used the flexibility of the ratio table to arrive at the answer differently. To demonstrate how the students used the ratio table to divide fractions, an example of one student's response to the number of cups needed of Corn Crunch 2 is shown in figure 24.2.

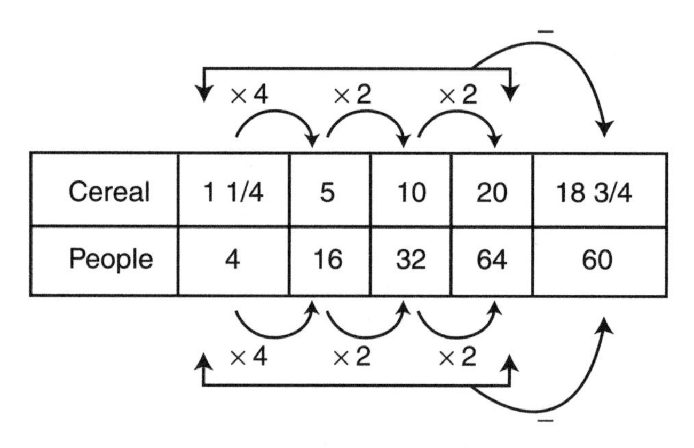

Cereal	1 1/4	5	10	20	18 3/4
People	4	16	32	64	60

Fig. 24.2. One student's use of the ratio table to divide fractions

This student explained that she had multiplied the terms of the ratio—1 1/4 cups of cereal to 4 people—by 4 to get the first equivalent ratio: 5 cups of cereal to 16 people. She then went on to explain that she had multiplied the the terms of the next two ratios by 2 to get the equivalent ratios 10:32 and 20:64. The student then explained that she knew she had to find the number of cups of cereal for 60 people. She stated that she had ratios with 64 people and 4 people, so she could subtract the similar quantities in these two equivalent ratios. Doing so left her with the ratio 18 3/4 cups to 60 people.

CONCLUSION

Throughout the rest of the section, Mrs. Pitingoro had the students present the problems and explain how they solved them until she was satisfied that they understood how to use the ratio table to make sense of and solve problems involving the division of fractions. Mrs. Pitingoro accomplished her goal: the students were using the ratio table to divide fractions. The ratio table gave many of Mrs. Pitingoro's students another way to divide fractions.

Listening to the students in both small-group and whole-class discussions made it apparent that the students' understanding of fractions had grown.

Mrs. Pitingoro taught in a manner envisioned in the *Curriculum and Evaluation Standards for School Mathematics* (NCTM 1989). The problems she used were open enough for students to reason about and communicate their strategies. She encouraged the students to explore dividing fractions while learning how to use the ratio table. The ratio table is not a pre-scripted, step-by-step algorithm. Instead, it is a flexible method of finding equivalent ratios that can be used for different purposes—among them, dividing fractions. Besides having different purposes, the ratio table is useful in that it allows students to be more cognizant of what they are solving and in that its use demonstrates how students are thinking about a problem.

REFERENCES

Encyclopaedia Britannica Educational Corp. *Mathematics in Context: Cereal Numbers.* Chicago, Ill.: Encyclopaedia Britannica Educational Corp., 1997.

Middleton, James A., and Marja van den Heuvel-Panhuizen. "The Ratio Table." *Mathematics Teaching in the Middle School* 1 (January-March 1995): 282–88.

National Council of Teachers of Mathematics. *Curriculum and Evaluation Standards for School Mathematics.* Reston, Va.: National Council of Teachers of Mathematics, 1989.

25

Teaching Statistics
What's Average?

Susan N. Friel

LEARNING statistics in the elementary and middle grades involves building both conceptual and procedural knowledge. From a conceptual standpoint, students need to develop "data sense." Data sense involves being comfortable with posing questions, collecting and analyzing data, and interpreting the results to respond to questions asked. It also includes being comfortable with—*and* competent in—reading, listening to, and evaluating reports based on statistics, such as those often found in the popular press. Further, data sense includes being able to make decisions about the kinds of analyses that are needed and, from a procedural standpoint, being able to use a variety of skills (e.g., compute statistics, create graphs) to complete these analyses. It is here that we find one direct application of a number of algorithms: those related to making tables, diagrams, and graphs, and those related to computing various statistics. The choices made as part of the analysis affect the interpretation and communication of the results. Developing conceptual and procedural understandings of how to construct the representations (e.g., tables, graphs, diagrams) or compute the statistics (e.g., median, mean, range) needed to analyze the data is important. With such understandings, students come to know what these representations and statistics do and how they behave in relation to data. Building data sense includes a consideration of the "when and why" of the use of the representations and statistics—a process of decision making that uses algorithms but clearly is not procedural itself.

BUILDING UNDERSTANDING OF MEASURES OF CENTER

Measures of center—mode, median, and mean—are summary statistics used for a variety of purposes, including to help describe or compare data sets. Almost any presentation of data uses averages; the mode, median, and

The author gives special thanks to Glenda Lappan for her comments on drafts of this paper.

mean are different ways to think about the broader concept of average. The *mode* is the most frequently occurring value in a set of data; it is any value having the highest frequency. The *median* is the midpoint of a set of data when items are arranged in order from least to greatest. The *mean* is the arithmetic average; it is the sum of the data values divided by the number of values. Conceptually, it can be viewed as a balance point of a distribution, operating like the fulcrum of a balance beam. One of the ways to describe what is typical about a set of data is to use one or more measures of center; generally median and mean are most useful for numerical data.

The importance of students' having strong conceptual understandings of these concepts cannot be overstressed; the value in their use comes from understanding what these measures "tell" about data. Yet, each is a "computation"; we can describe algorithms for determining the mode, the median, and the mean for a set of data. There is an interaction between building conceptual and procedural understandings that needs to address both what each measure "means" and how the algorithms may evolve through a development of conceptual understanding.

The Mode

The mode is the simplest measure of center to "compute." It is the value that has the "most Xs" on a line plot or "the tallest bar" on a bar graph when the data are shown on a graph (see "Numbers of Pets" in fig. 25.1). However, its meaning may easily be misstated, that is, *most students in the class have two pets in their households* versus *households with two pets occur most often in this set of data.* The latter statement is correct and exhibits a subtle, yet important, distinction in language. Focusing students' attention on actual frequencies helps highlight this distinction, that is, only about 19 percent, or one-fifth, of the students in the class have two pets. The mode also is not a stable statistic; its value fluctuates easily with changes in the data. If another student joining the class had three pets, the data would be bimodal (two modes). Or if two students, each having three pets, joined the class, the mode would change from two pets to three pets.

If we think about the kinds of data we may be using (categorical or numerical), the mode is the only measure of center that can be used to describe categorical data (e.g., in "Favorite Kinds of Pets" in fig. 25.1, "dog" occurs most often as the choice for favorite kind of pet; about 27 percent of the students chose dogs as their favorite pet). The mode is not a particularly useful statistic to use with numerical data when the other two measures are available.

The Median

There are different ways to build conceptual understanding of the median as a measure of center and, from that, to develop algorithms. In an ordered set of data, the median is the value of the middle item of data (if there is an odd number of data values) or the average of the two middle items of data (if there is an even number of data values). One way to locate the median is to list the data values (e.g., numbers of pets, lengths of names, or heights) in order from smallest to largest and to count in from the ends by pairing one value from the "small" end with one value from the "large" end until the middle is reached. Another way (using lengths of names) is to record the lengths in order from smallest to largest on a strip of one-inch (or other suitable size) grid paper, one length on each square grid (Lappan et al. 1996). The strip is folded in half to identify the median (see fig. 25.2).

Favorite Kinds of Pets

Pet	Frequency
cat	4
dog	7
fish	2
bird	2
horse	3
goat	1
cow	2
rabbit	3
duck	1
pig	1

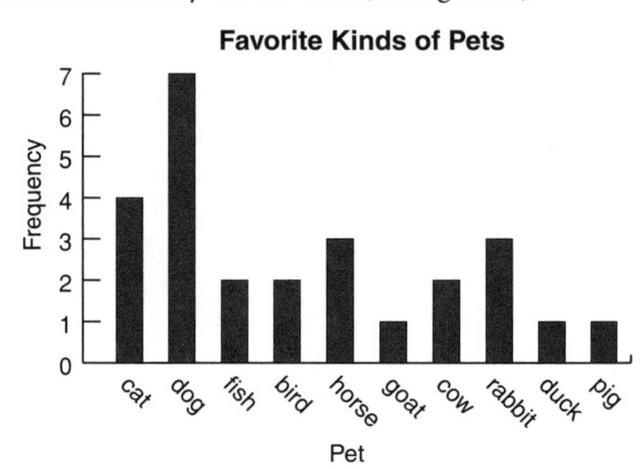

Numbers of Pets

Number of pets	Frequency
0	2
1	2
2	5
3	4
4	1
5	2
6	3
7	0
8	1
9	1
10	0
11	0
12	1
13	0
14	1
15	0
16	0
17	1
18	0
19	1
20	0
21	1

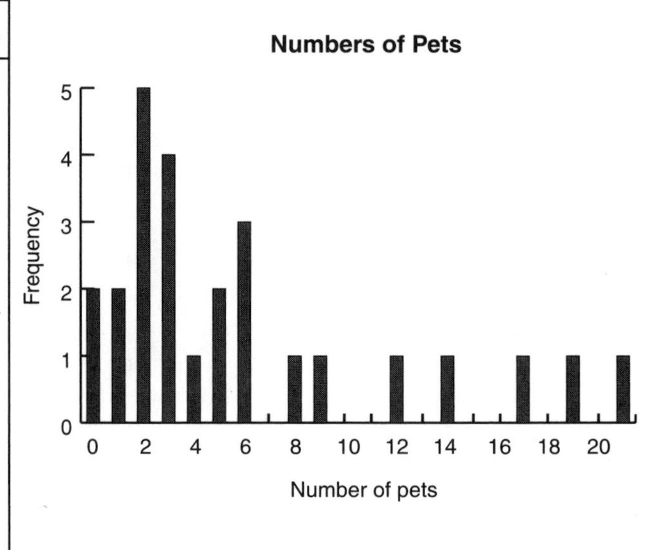

Fig. 25.1. Illustrating the mode (reprinted from Lappan et al. 1996, p. 24)

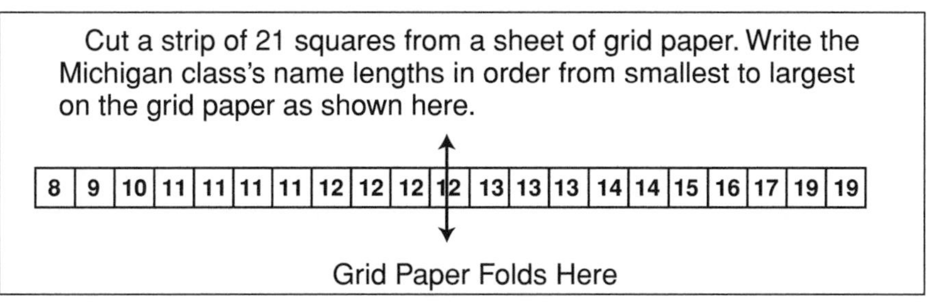

Fig. 25.2. A data strip can be folded in half to identify the median (reprinted form Lappan et al. 1996, p. 12).

A teacher may find that other opportunities are needed for students to work with the median with small sets of data before going on to a discussion that focuses on finding the median of a large data set. To foster the development of algorithms to use with larger data sets, students may consider what would happen if they had lengths of names for thirty students. In such a context, students realize that the data have to be in order. Some students want to find the "halfway point." They do so by halving the total number of data values in the data set and determining that halfway is between the fifteenth and sixteenth data values. Many use the conceptual model to help them visualize these data listed in order on a strip of grid paper with a fold at the middle of the strip.

To help clarify and refine their procedures, students are encouraged to think about situations in which there are fifty or seventy-five names. The size of the data set is used to impress on students the need for fine-tuning strategies so they have efficient algorithms for identifying the median in a given data set. Many students realize that when the data set has an even number of values, dividing the number of values in half tells how many values to "count in" from one end of the ordered data set. The median is halfway between this data value and the next data value. When the data set has an odd number of values, dividing the number of values in half results in a fraction; this fraction is rounded to the next whole number, which tells which value in the data set is the median. Formally, if there are n data values, the algorithm involves arranging the values in order from least to greatest and counting in $(n + 1)/2$ values from the least value to find the location of the median.

Students can also explore the *stability* of the median as a statistic. Suppose that they have a list of the lengths of students' names and a new student joins the class whose name is longer than any other student's name. Adding the length of this name to the set of data will result in the same change in the median no matter what the length of the name is.

The Mean

There are two general strategies (Friel, Mokros, and Russell 1991; Lappan et al. 1996; Mokros and Russell 1995; Russell and Mokros 1996) for helping

Data Value	Mean	Difference between Value and Mean	
0	6	6	
0	6	6	
1	6	5	
1	6	5	
2	6	4	
2	6	4	
2	6	4	
2	6	4	
2	6	4	58
3	6	3	
3	6	3	
3	6	3	
3	6	3	
4	6	2	
5	6	1	
5	6	1	
6	**6**	**0**	
6	**6**	**0**	
6	**6**	**0**	
8	6	2	
9	6	3	
12	6	6	
14	6	8	58
17	6	11	
19	6	13	
21	6	15	

students visualize the mean: using the "fair share" or "evening out" model or using the balance model. In the fair-share model, if each student in a class has a certain number of pets (including 0 pets), then we can combine all the pets and redistribute the numbers equally among all the students in the class (including those who had 0 pets). The number of pets each student then has is the mean (6 pets).

In the balance model, the mean is a value about which the data "balance." Unlike the median, which has the same number of data values on either side, the sums of the differences between the data values and the mean on either side of the mean are equal. Continuing with the example involving the number of pets, figure 25.3 shows the difference from the mean in the number of pets for each student.

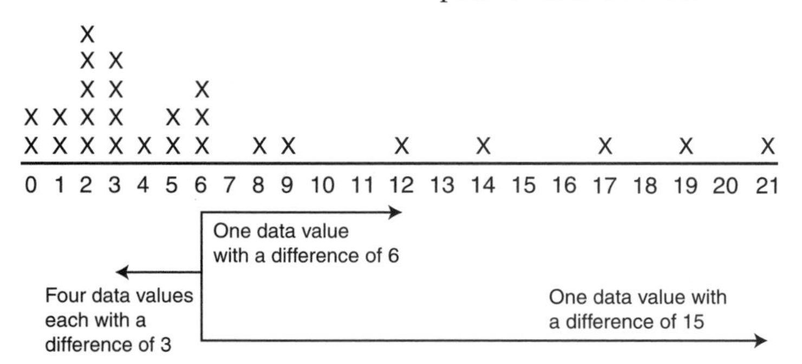

Fig. 25.3. In this balance model, the sum of the differences "below" the mean is 58, and the sum of the differences "above" the mean is 58. The line plot above demonstrates this concept.

Research has shown that either model can be used to help students understand the concept of mean (Mokros and Russell 1995). In our work in the Connected Mathematics Program, we found that teachers with little content background in statistics and little preparatory professional development in teaching statistics were more successful in presenting a visual model of the mean developed through the fair-share or evening-out model. In addition, the evening-out model lends itself to the development of the standard algorithm for finding the mean. In this model, the conceptual view of mean as a balance point is noted but not emphasized.

Some introductory work that develops the evening-out model is shown in figure 25.4. Once a representation is made with cubes, students are asked to make a line plot showing the same data using stick-on notes to display the frequencies for each size of household. It is not always obvious to students that a stick-on note shown above a 6 on a line plot means that there are six people in that household as shown by a cube tower. Once the tower is "collapsed" into a stick-on note and the same kind of stick-on note represents households of different sizes, students can become confused. Explicitly selecting a stick-on note and asking students what value it represents is one way to help students focus on this distinction.

To build conceptual understanding, students explore making data sets that have different means. In working with colored cubes, it may be possible to make each tower a different color. When the data are "evened out," the colors of the cubes help students visualize the redistribution. The colors also make it possible to work backward and recreate the original data set. Such an activity helps students isolate the components that are related to finding the mean of a set of data: the number of data values, the sum of the data values, and the computed mean. Giving words to their actions, the students conclude that the mean is the number of people in a household if everyone had the same number of people in the household. Again, students do not necessarily use these words, but their discussion points to a developing understanding of this definition.

As with the median, larger data sets are introduced and students are asked to discover ways to find the mean (i.e., algorithms) without relying on cube towers or line-plot representations. Three components are important in developing the algorithm: the number of data values, the sum of the data values, and the computed mean. Any one of these values may be the unknown; usually students are asked to compute the mean when they know the number of data values and the sum of the data values. The algorithm—find the sum of the data values and divide by the number of data values—has a referent in earlier experiences with the cube towers.

In addition, students can explore the stability of the mean as a statistic. We return to an earlier example: if students have a list of the lengths of students' names and a new student joins the class whose name is longer than any other student's name, the length of that name *does* matter when determining the mean. Unusually large or small data values have the effect of "pulling the mean" higher or lower.

USING MEASURES OF CENTER

Developing conceptual knowledge about what these three measures of center tell us and why we use them requires that students engage in problems in

Evening Things Out

Eight students in a middle-school class determined the number of people in their households using the United States census guidelines. Each student then made a cube tower to show the number of people in his or her household.

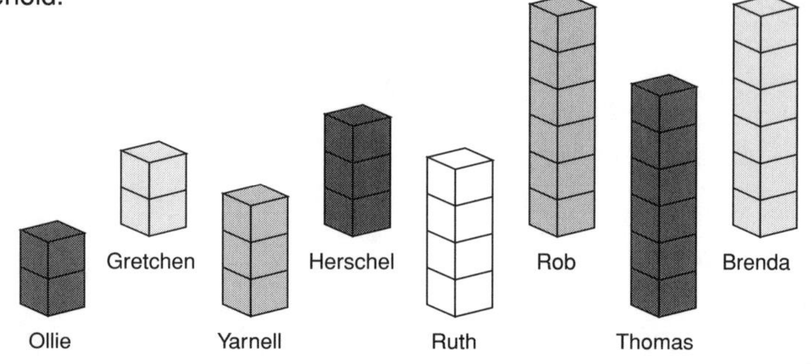

You can easily see from the cube towers that the eight households vary in size. The students wondered what the average number of people is in their households. Their teacher asked them what they might do, using their cube towers, to find the answer to their question.

Problem 5.1

What are some ways to determine the average number of people in these eight households?

Problem 5.1 Follow-Up

The students had an idea for finding the average number of people in the households. They decided to rearrange the cubes to try to "even out" the number of cubes in each tower. Try this on your own and see what you find for the average number of people in the households, and then read on to see what the students did.

First, the students put the towers in order.

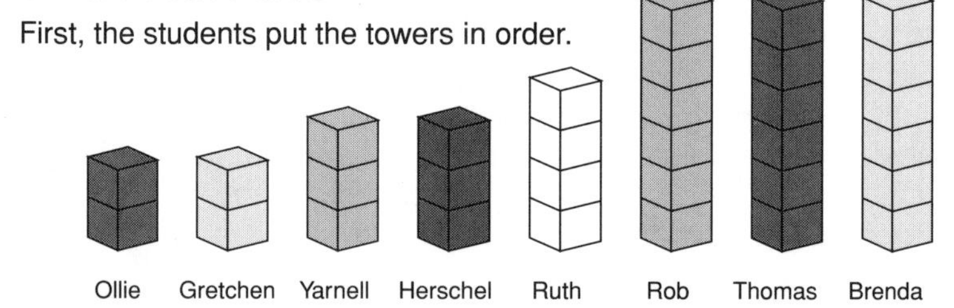

The students then moved cubes from one tower to another, making some households bigger than they actually were and making other households smaller than they actually were. When they were finished moving cubes, their towers looked like this:

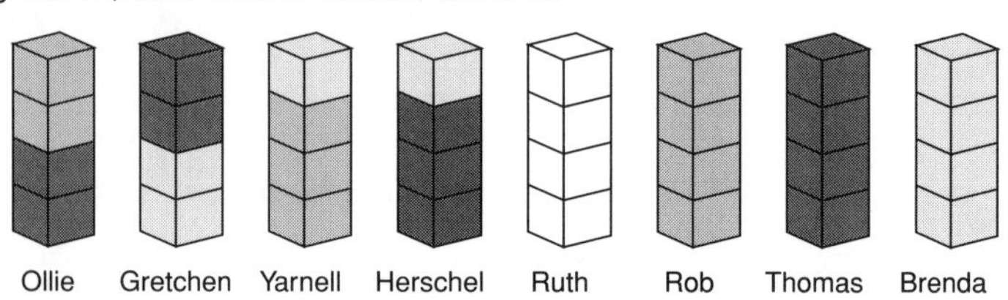

Each tower now had four cubes. Notice that the total number of cubes did not change.

Before

Ollie	2 people
Gretchen	2 people
Yarnell	3 people
Herschel	3 people
Ruth	4 people
Rob	6 people
Thomas	6 people
Brenda	6 people
Total	**32 people**

After

Ollie	4 people
Gretchen	4 people
Yarnell	4 people
Herschel	4 people
Ruth	4 people
Rob	4 people
Thomas	4 people
Brenda	4 people
Total	**32 people**

The students determined that the average number of people in a household was 4. The teacher explained that the average the students had found is called the **mean.** The mean number of people in the eight households is 4.

Fig. 25.4. One way to develop the evening-out model (reprinted from Lappan et al. 1996, pp. 54–55)

a number of different contexts. The more situations arise in which students are asked to distinguish among the three measures and, given the context, make decisions about which to use, the greater students' understanding that the measures of center are more than just numbers to be computed whenever they have data. As an example, figure 25.5 presents a problem in which students rely on their understandings of the characteristics of the measures of center to determine which measure is being used.

More-complex situations require assessing the best way to present a summary of the data. For example, local newspapers often conduct surveys of their readers. One newspaper conducted a poll to find out how their comic strips rated with different groups of readers. The poll asked readers to rate each of the comic strips on a scale of 1 to 5. Once the data are gathered, the goal is to describe readers' preferences. The ratings for each comic strip need to be summarized. Is it better to give mode, mean, or median ratings for each strip? What about reporting percents for each number on the scale? To answer these questions requires that students think about (1) what each measure tells them about the data and (2) how this information will help in the final stage of interpreting and communicating the results.

CONCLUSION

The mode, median, and mean are kinds of averages that are part of a tool kit of representations and statistics used to analyze data. Ideally, we want students to understand what each of these measures tells about a set of data,

8. Three candidates are running for the mayor of Slugville. Each has determined the typical income for the people in Slugville and is using this information to help in the campaign.

Mayor Phibbs is running for re-election. He says, "Slugville is doing great! The average income for each person is $2000 per week!"

Challenging candidate Louisa Louis says, "Slugville is nice, but it needs my help! The average income is only $100 per week."

Radical Ronnie Radford says, "No way! We must burn down the town—it's awful. The average income is $0 per week."

None of the candidates is lying. Slugville has only 16 residents, and their weekly incomes are $0, $0, $0, $0, $0, $0, $0, $0, $200, $200, $200, $200, $200, $200, $200, and $30,600.

a. Explain how each of the candidates determined what the "average" income was for the town. Check the computations to see whether you agree with the three candidates.

b. Does any person in Slugville have the mean income? Explain.

c. Does any person in Slugville have the median income? Explain.

d. Does any person in Slugville have the mode income? Explain.

e. What do you consider to be the typical income for a resident of Slugville? Explain.

f. If four more people moved to Slugville, each with a weekly income of $200, how would the mean, median, and mode change?

Fig. 25.5. A problem that requires students to use their understandings of the characteristics of measures of center to determine which measure is used in different situations (reprinted from Lappan et al. 1996, p. 64)

make judgments about when and why to use each measure when they analyze one or more sets of data, and be comfortable in knowing how and why they carry out the actual algorithms. Such understanding develops over time and with increasing sophistication as students are given various problems in which to explore their thinking. Eventually, students move from conceptual to procedural understanding, able to use the necessary algorithms and to focus their attention on the interpretations of their results based on their understandings of what they know about data given various statistics and representations.

REFERENCES

Friel, Susan, Jan Mokros, and Susan Jo Russell. *Statistics: Middles, Means, and In-Betweens.* Palo Alto, Calif.: Dale Seymour Publications, 1991.

Lappan, Glenda, James Fey, William Fitzgerald, Susan Friel, and Elizabeth Phillips. *Data about Us*. Connected Mathematics Project. Palo Alto, Calif.: Dale Seymour Publications, 1996.

Mokros, Jan, and Susan Jo Russell. "Children's Conception of Average and Representativeness." *Journal for Research in Mathematics Education* 26 (1995): 20–39.

Russell, Susan Jo, and Jan Mokros. "What Do Children Understand about Average?" *Teaching Children Mathematics* 2 (1996): 360–64.

26

Algorithms for Solving Nonroutine Mathematical Problems

Jinfa Cai

John C. Moyer

Connie Laughlin

MATHEMATICAL algorithms are powerful tools that contribute to effective problem solving and are rules that guarantee a solution when correctly applied. However, there is a great deal of empirical evidence that although some students appear to know an algorithm, they cannot correctly apply the algorithm to solve a problem. Understanding an algorithm conceptually implies knowing the procedures specified by the algorithm and knowing when and how the procedures should be applied. The question is, How can teachers create learning environments so that students can understand where and when to use the algorithms? About a half-century ago, Piaget had already indicated that "to understand is to discover, or reconstruct by rediscovery, and such conditions must be complied with if in the future individuals are to be formed who are capable of production and creativity and not simply repetition" (Piaget 1973, p. 20). Today, the idea that students should be allowed to invent their own algorithms is still valued (National Council of Teachers of Mathematics [NCTM] 1989, 1991). "Mathematics is learned when learners engage in their own invention and impose their own sense of investigation and structure" (NCTM 1991, p. 144). Thus, in order to know when and how to apply algorithms correctly in problem solving, students should be provided opportunities to develop and invent their own algorithms.

In this paper, we share with readers examples in which students were given situations and guided to invent their own procedures and algorithms. Students actively participated in processes of knowledge construction and made sense of mathematical procedures and algorithms. They became active participants in creating knowledge rather than passive receivers of rules and procedures. The lesson plans were designed to allow the students to invent their own algorithms in solving nonroutine problems. The two examples presented here involve middle school students' inventing algorithms through inductive reasoning and generalization. The concepts and skills involved in these two examples are very important for the development of their algebraic thinking.

ALGORITHMS FOR FINDING THE SUM OF AN ARITHMETIC SERIES

As students move into the middle grades, the development of algebraic reasoning becomes more and more important. Students should be given many problems that allow them to study patterns and look for ways to generalize results. Eventually, these generalizations are expressed in closed algebraic forms using variables. One such problem that was posed to a group of sixth-grade students is the Staircase problem, shown in figure 26.1. The students were asked to find their answers in as many different ways as they could. Findings from cross-national studies show that teachers in Japan and China experienced success with this teaching strategy.

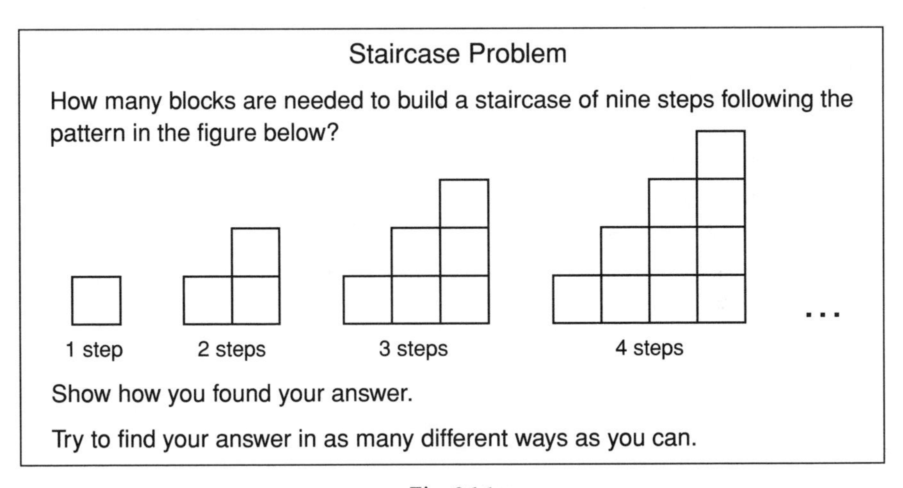

Fig. 26.1

The students worked on this problem in pairs. After twenty minutes, each pair had at least one way to determine the number of blocks needed to build a staircase of nine steps. Then each pair was asked to present its solutions. The initial strategies were dependent on the specific cases and unsophisicated. Most students focused on drawing all nine stair steps and then counting. Some students said that in order to find the total number of blocks needed to build a staircase of nine steps, they just found the sum of $1 + 2 + 3 + 4 + 5 + 6 + 7 + 8 + 9$. If the numbers are added in their original order, the result will be the total number of blocks. Many groups recognized that to find the total number of blocks needed in a staircase, they could just add the number of blocks in the new step to the previous total number of blocks. Of course, this method requires them to know the previous sum, so the students did not have an efficient way to get past this obstacle. Two invented algorithms were presented by two pairs and elaborated through classroom discussions: a pairing algorithm and a geometric algorithm. These two algorithms were developed simultaneously in the classroom, but they are presented separately in this paper.

The Pairing Algorithm

Frank and Maria looked for ways to add the numbers other than in the original order and soon discovered the idea of pairing numbers to obtain common sums:

$$\begin{aligned} 1 + 2 + 3 + 4 + 5 + 6 + 7 + 8 + 9 &= (1 + 9) + (2 + 8) + (3 + 7) + (4 + 6) + 5 \\ &= 10 + 10 + 10 + 10 + 5 \\ &= 10 \times 4 + 5 \\ &= 45 \end{aligned}$$

The symmetry of this approach appealed to many pairs, but they did not accept it as a superior solution strategy until the problem was extended to a greater number of stairs.

After all the pairs had presented their solutions, the teacher asked the students, "What if we need to build a staircase of fifty steps, how many blocks do we need?" After a pause, the teacher continued, "We have already discussed several ways to figure out the number of blocks needed to build a staircase of nine steps. What if we used these ways to find out the number of blocks we need to build a staircase of fifty steps?" The earlier algorithms of just counting by ones or adding in order were discarded by many students as too inefficient. Instead, this pairing algorithm became very popular. Jane in particular was very proud to share her solution to this problem with the class. Her work is shown in figure 26.2.

After Jane's presentation, many students were puzzled because Jane's answer was different from their answer. Some other students were puzzled because many of Jane's pairs were not equal to 51. Jane had paired 25 with 27, 24 with 28, and so forth, but the sum of each of these pairs is 52, not 51 as Jane had said. Cognitive conflict leads to many constructive discussions in class. The teacher asked the students to investigate what had happened in Jane's solution. Bill raised his hand and pointed out that when Jane had done the pairing, she had skipped 6. Bill then offered his correct pairing algorithm: "There are 25 pairs of 51. $51 \times 25 = 1275$. So the correct answer should be 1275." Some students also agreed with Bill that 1275 should be the correct answer.

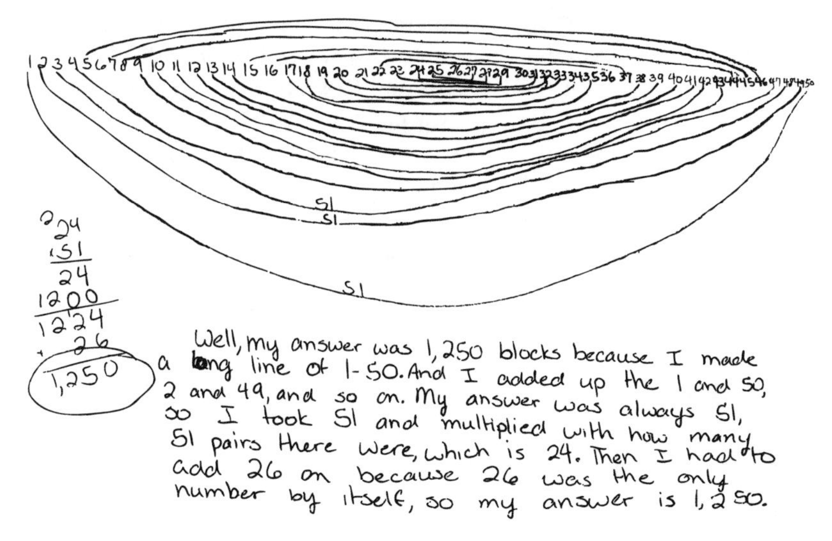

Fig. 26.2. Jane's pairing algorithm for the Staircase problem

When Bill correctly paired the numbers, he did not have an unpaired middle number. That seemed to puzzle Jane. She asked, "Where is the middle one, which is not paired with anybody?" Most of the students and the teacher in the class seemed to have difficulty understanding what Jane meant. The teacher asked, "What do you mean, 'the middle number,' Jane?" Jane replied, "Well, if you look at nine steps, you paired the numbers, but the middle number 5 is not paired with anybody." Then the teacher understood what Jane meant. Since with nine steps there is a middle number, 5, that is not paired, Jane appeared to believe that there should also be an unpaired middle number with 50 steps. So the teacher focused the class discussion on the resolution of Jane's question.

- What is the common sum?
- How many pairs do you have? How do you know?
- Is any number unpaired? How do you know?
- What happens when pairing with even numbers of steps and odd numbers of steps?

The discussion ended at the point when the students realized that if they have an even number of steps, they will not have an unpaired middle number and if they have an odd number of steps, they will have an unpaired middle number.

The Geometric Algorithm

On the first day of exploring this problem, Bob and Tasha suggested drawing a complete 9-by-9 square and then "slicing it in half" to find the number of blocks in the nine-step staircase (see example 1 in fig. 26.3). These two students knew they were on to something but could not fully explain it to the class at the time. When this picture was shared with the whole class, many interesting questions were raised by the other students, for instance, "Was the square really divided in half? Half of 81 is not 45, is it?"

Example1: Bob and Tasha's algorithm group

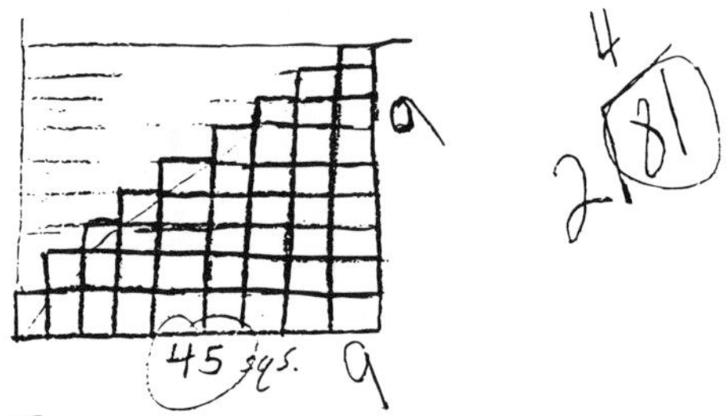

Example 2: An algorithm from Patti's group

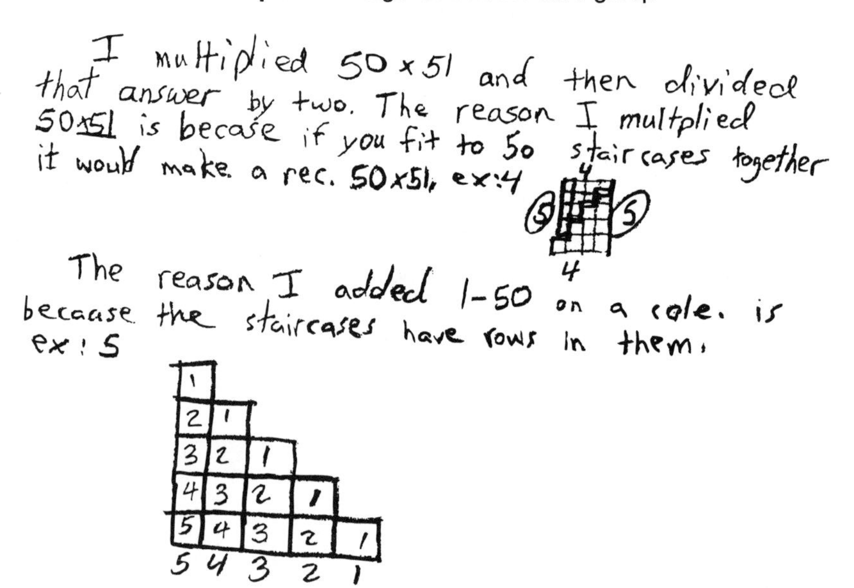

Fig. 26.3. Two students' geometric algorithms for the Staircase problem

Later, the teacher helped the students in the class see that the square was not actually divided in half by encouraging them to draw a diagonal line on Bob and Tasha's figure. All the students agreed that the diagonal line divided the 9-by-9 square in half but that it cut off some of the staircase. A close examination of the figure shows that the nine-step staircase is more than half of a 9-by-9 square, since there are nine half-squares above the diagonal. The 9-by-9 square has 81 blocks, and half is 40.5. If we add the 9 half-blocks to 40.5, we will get 45, the number of blocks in a nine-step staircase. This classroom discussion helped the students clarify ideas about shape, geometry vocabulary, and division.

Patti also presented her group's solution to the class. Her solution was similar to Bob and Tasha's, but she fit two nine-step staircases together, which formed a rectangle rather than a square. Example 2 in figure 26.3 shows her group's solution. This solution was the catalyst for many students to begin to generalize the problem. If the staircase had ten steps, the rectangle was 10×11. Patti explained, "Then you divide the product of 10×11 in half because you only want one staircase, not two." If the staircase had fifty steps, the rectangle formed by two staircases was 50×51. Several groups used this idea as one solution to a fifty-step staircase. It was interesting to watch them arrive at answers to various sizes of staircases. They used the pairing algorithm and the drawing strategy together with counting ideas to arrive at their answers, and they also began to express a generalization for the problem of adding consecutive integers.

After three days of working on this problem, many groups were able to begin to generalize in words about how to obtain the number of steps in any staircase. The teacher decided to go one step further and encouraged the students to use variables to show the number of blocks in any staircase. A few students wrote the generalization in symbols. When they shared this formula with the rest of the class, the other students listened but could not readily apply it to new staircases. For the most part, the majority of the students fell back on their earlier algorithms for help. They either (1) needed to list a few, if not all, of the numbers in pairs to obtain the sum or (2) needed to draw a rectangle composed of two sets of staircases. Their reliance on their earlier algorithms suggests that their algorithmic thinking about the sum of consecutive numbers or the area of a rectangle is a facilitating step on the road to understanding the formula for the sum of consecutive integers. As these students encounter this same number situation in new problems, they will have at least one or two fairly efficient strategies to obtain the sum. Eventually, most may use the formula, but these intermediate strategies are the key to making sense of the ideas about symbolism.

ALGORITHMS IN THE CROSSING-THE-RIVER PROBLEM

A certain aspect of algorithms can be described by numeric patterns that, in turn, can be modeled with algebraic symbols. In this sense, algebraic reasoning is at the heart of algorithmic processes. The second example shows how algorithms invented by students for solving a problem can also stimulate algebraic thinking. Given an appropriate problem, students can be induced to conceive a strategy for solving the problem, invent an algorithm for applying the strategy, and model the algorithm with algebraic symbols. Teachers who observe their students engaging in problem solving of this type have a golden opportunity to assess their students' level of algebraic thinking.

The problem that we have used with middle school students is Crossing the River, shown in figure 26.4. The students worked in groups of three. Almost all the students found this problem to be enjoyable and interesting. They soon became enthusiastically immersed in the solution process. At first the students did not have a thorough understanding of the problem. In order to understand it, the groups needed to act out various scenarios with manipulative materials. Through trial and error, the students soon understood the crux of the problem and began to invent and formulate a strategy for solving it. At the heart of their strategy was an awareness that when an adult crosses to the far side of the river, there must always be a child waiting there to bring the boat back.

Crossing-the-River Problem

Eight adults and two children need to cross a river. A small boat is available that can hold one adult or one or two children. Everyone can row the boat.

What is the minimum number of trips needed for all the adults and children to cross the river?

Show or explain how you got your answer.

Fig. 26.4

From Strategy to Procedure

A realization of the basic strategy alone, however, was not sufficient to solve the problem. The students had to convert this basic strategy into a procedure for getting all eight adults and two children across the river in the minimum number of trips. In time, this procedure led to the invention of a generalized algorithm for solving all problems of this type. It was interesting to observe the students as they grappled to "proceduralize" their strategy (i.e., to develop an algorithm). Many, even after they had conceptualized a correct procedure, had difficulty implementing it without error. For some, colored counters were too abstract. They had to resort to the use of more-realistic materials: a model that looked like a boat, large pieces for adults, and small pieces for children.

After a while, all the groups came up with an answer. Most of the answers were correct: thirty-three trips. Wrong answers were not due to faulty strategies. Rather, wrong answers were due to incorrect counting or incorrect compensation for an error in implementation. Through discussions with the students about their procedures, the teacher realized that many groups were not yet explicitly aware of the repetitive sequence of steps they had used to solve the problem. This meant that even though the students were able to "proceduralize" their strategy for solving the problem, the procedure itself was not yet sufficiently understood at a conscious level. The teacher's goal was for the students to transform their procedure into a generalized concrete algorithm. To do so, the students needed to conceptualize the steps individually as well as in sequence with other steps.

From Procedure to Algorithm

To induce the students to make a conscious effort to invent a generalized algorithm, the teacher then asked the students to record their solutions to three different variations of the problem: to find the number of one-way trips it takes for (1) six adults and two children, (2) fifteen adults and two children, and (3) three adults and two children to cross the river. Note that the teacher changed only the number of adults, not the number of children. With the number of children maintained at two, the solution for each situation varies only in length. The teacher told the students to devise their own way of recording their solution and to record it while the pieces were actually manipulated. He told them that one student should describe how a second student is manipulating the pieces while a third student records the information. This requirement not only ensured that all three students were involved in the process, but it also provided tactile, auditory, and written feedback on the repetitive nature of the procedure.

Through this process, the students were led to invent a generalized algorithm that they could use to solve any problem of this type. At the heart of the algorithms invented by the students was a "chunk" of four moves that was repeated for each adult transported to the far side. The groups became aware of the repetition of chunks in different ways. Akeem, Jorge, and Tyra's group became very enamored with the rhythm of the words that Tyra kept repeating for Akeem to write down: "Two kids, one kid, one adult, one kid; two kids, one kid, one adult, one kid; two kids, one kid, one adult, one kid;" and so on. Sam, Rosa, and Tara noticed the chunks by examining the diagrams they had made. So did Ning, DeAnn, and Thomas. The last two groups represented the elements of the pattern in similar ways, but the elements appeared in the chunks in a different order. Examples 1 and 2 in figure 26.5 show the difference between their algorithms.

Sam, Rosa, and Tara noticed chunks that started at the top of their diagram. In their algorithm, a chunk is composed of the first four crossings: two children cross to the far side, one child crosses back to the near side, one adult crosses to the far side, one child crosses back to the near side. One

Chunk	Pattern
Chunk 1 gets first adult across	2 children ⟶ ⟵ 1 child 1 child ⟶ ⟵ 1 child
Chunk 2 gets second adult across	2 children ⟶ ⟵ 1 child 1 adult ⟶ ⟵ 1 child
Chunk 3 gets third adult across	2 children ⟶ ⟵ 1 child 1 adult ⟶ ⟵ 1 child
Final trip	2 children ⟶

Example 1: Sam, Rosa, and Tara's conceptualization of the chunks in their algorithm

Pattern	Chunk
2 children ⟶	Startup trip
⟵ 1 child 1 adult ⟶ ⟵ 1 child 2 children ⟶	Chunk 1 gets first adult across
⟵ 1 child 1 adult ⟶ ⟵ 1 child 2 children ⟶	Chunk 2 gets second adult across
⟵ 1 child 1 adult ⟶ ⟵ 1 child 2 children ⟶	Chunk 3 gets third adult across

Example 2: Ning, DeAnn, and Thomas's conceptualization of the chunks in their algorithm

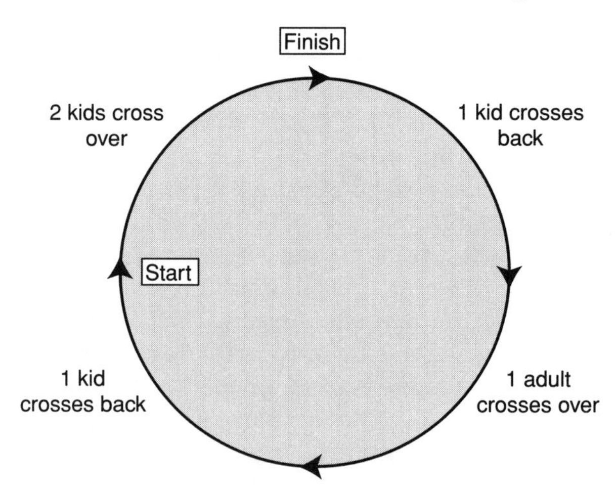

Example 3: A cyclic algorithm from Terry, Glenn, and Roberta's group

Fig. 26.5. Algorithms for the Crossing the River problem

chunk must be performed for each adult who crosses the river. These chunks are repeated until only two children are left on the near side. Finally, one extra trip is required to complete the solution.

Ning, DeAnn, and Thomas noticed chunks that occurred at the bottom of their diagram. In their algorithm, the last four crossings constitute a chunk that had been repeated throughout their solution: one child crosses back to the near side, one adult crosses to the far side, one child crosses back to the near side, two children cross to the far side. Ning, DeAnn, and Thomas's algorithm requires a single trip at the beginning before the repetitive chunking begins: two children must cross to the far side, then one chunk is performed for each adult who crosses the river.

Joe, Enrique, and Cybill decided to draw arrows of different colors to represent different combinations of people crossing the river. In their algorithm, a red arrow represents two children crossing, a green arrow represents a single child, and a blue arrow represents a single adult. This group identified the chunks through the repetition of the colors: red, green, blue, green; red, green, blue, green; red, green, blue, green; red, green, blue, green; and so on.

Terry, Glenn, and Roberta conceptualized the chunks as cyclic. They drew an insightful diagram (see fig. 26.5, example 3), in which the chunks are repeated cyclically until only two children remain on the near side. Then the final trip to get the two children across is simply the first leg of the next cycle.

By the time the students finished solving these problems, they were able to implement their strategy in a repetitive procedure with few errors. In addition, by forcing the students to attend to the details of the strategy, the teacher guided them in inventing a generalized algorithm that they could apply almost flawlessly to any Crossing the River situation with two children.

From Algorithm to Symbolic Representation

The teacher's final goal in this process was to induce the students to model the algorithm algebraically. To do so, the teacher asked the students, "Can you describe, in words, how to figure out the answer for this problem if the group of people to cross the river includes 2 children and any number of adults? How does your rule work out for 100 adults?" The teacher used such a large number of adults to encourage the students to think of a way to solve the problem other than by manipulating or drawing each trip. Finally, the teacher challenged the students, "Can you generalize the rule for 2 children and any number of adults, say, x?"

Tyra quickly started chanting, "Two kids, one kid, one adult, one kid; two kids, one kid, one adult, one kid; two kids, one kid, one adult, one kid..." while Akeem counted the number of crossings and Jorge counted the number of adults. Pretty soon, however, they got confused. Then Akeem said, "Each adult needs 4 trips, so that's 400 trips. We don't need to count them. It's four times the adults." He forgot about the extra trip at the end, and so did Tyra and Jorge. To help them remember, the teacher suggested they compare their solution with Joe, Enrique, and Cybill's solution.

Joe, Enrique, and Cybill had not forgotten about the extra trip. They looked at their colored diagrams and were able to write, "Four trips for every adult, plus the last trip for the 2 kids. $4 \times 100 + 1$. $4x + 1$." Ning, DeAnn, and Thomas also were quickly able to solve the problem, writing "$1 + 4 \times 100$. The first trip to get the two kids across and then 4 trips for every adult. $1 + 4x$."

Sam, Rosa, and Tara, though, took a different approach. They concentrated on the directions of the arrows in their diagram (fig. 26.5, example 1) and separated them into two types: those going over and those coming back. They explained, "Each adult needs 2 trips over and 2 trips back, and we can't forget the last trip for two kids. So, you take the number of adults and multiply by 2. That is, $100 \times 2 = 200$. You add 1 for the last trip = 201. Then you add $200 + 201 = 401$." The general form of the solution the students provided was $2a + 2a + 1$ (here a stands for the number of adults).

One student, José, had a very different approach to this problem. He used the results of prior problems and some proportional reasoning to arrive at his answer. His group had already determined that with two children, three adults require thirteen trips, fifteen adults require sixty-one trips, and six adults require twenty-five trips. So he reasoned that the first twenty-four adults ($3 + 15 + 6$) to cross would require ninety-six trips ($13 + 61 + 25 - 3$), since the total of the trips from prior problems was ninety-nine but the three extra trips must be subtracted because the three groups are separate. Next he multiplied the 24 adults and the 96 trips each by 4 to determine first that 96 (i.e., 24×4) adults would require 384 (i.e., 96×4) trips. He then separately figured out that he needed 17 more trips to get the remaining 4 adults and 2 kids across. So his final answer was 401. He tried to write a rule for x adults and 2 children but got frustrated and gave up. The teacher mentally noted that although José's solution was quite creative, she would watch him in his discussions with other students to see if he could adopt an approach more conducive to symbolic representation.

The examples given here show the importance of having students invent their own algorithms. It is clear that each group of students had its own unique way of conceptualizing the Crossing the River problem. Because the teacher encouraged the students to invent their own algorithms, they were given an avenue to express their individual conceptualizations of the problem. Being able to conceptualize the problem indicated that the students really understood the generalizations they eventually discovered. More important, the teacher avoided the pitfall of imposing a conceptualization of the problem and a generalization that the students did not understand. The discussions of the students' solutions to the problem naturally lead them to make sense of using variables to represent patterns.

CONCLUSION

As Steen (1986, p. 6) indicates, "the mathematics curriculum must not give the impression that mathematical and quantitative ideas are the product of authority or wizardry." Students have "authority" over the algorithms through their own inventing and their constructive processes in solving nonroutine problems. When students invent algorithms, they will not only know the algorithms but will also know when and how to apply the algorithms correctly to novel situations. Students should be encouraged and allowed such invention and knowledge-construction processes in the classroom. Moreover, mathematics instruction at the middle school level is essential to preparing students for their study of algebra in high school. The development of their algorithmic thinking is an important component of the preparation.

REFERENCES

Kamii, Constance. *Young Children Continue to Reinvent Arithmetic (Third Grade): Implications of Piaget's Theory.* New York: Teachers College Press, 1994.

National Council of Teachers of Mathematics. *Curriculum and Evaluation Standards for School Mathematics.* Reston, Va.: National Council of Teachers of Mathematics, 1989.

———. *Professional Standards for Teaching Mathematics.* Reston, Va.: National Council of Teachers of Mathematics, 1991.

Piaget, Jean. *To Understand Is to Invent: The Future of Education.* Translated by George-Anne Roberts. New York: Grossman Publishers, 1973.

Steen, Lynn. "A Time of Transition: Mathematics for the Middle Grades." In *A Change in Emphasis,* edited by Richard Lodholz, pp. 1–9. Parkway, Mo.: Parkway School District, 1986.

Part 5

27

Algebra and Technology

Ann Brunner

Kathy Coskey

Sharon K. Sheehan

IN THIS paper we examine three uses of technology that enable students to develop new algorithms for solving algebra problems. These new approaches allow different interpretations of "to solve," all of which are based on a function approach to applications. The problems include—

- solving for a zero of a function from a graph;
- solving an inequality by using a table or a graph;
- solving a linear equation by using a table;
- solving a system of equations by using a table.

The technology used includes a TI-82 graphing calculator, the symbolic computational software program Mathematica (1993), and spreadsheets. However, other graphing calculators and software packages can be used in similar ways.

SOLVING FOR A ZERO OF A FUNCTION FROM A GRAPH

The following problem comes from Fey and Heid (1991). The students are already familiar with the graphing calculator for defining functions, graphing, and making tables. They also know how to identify the independent and the dependent variables.

> *Situation:* Two girls want to enter the summer lawn-mowing business. They plan to buy a mower for $140. They hope to earn $8 per hour of work. Their profit in dollars will be a function of time worked in hours, with the rule $y = 8x - 140$, where y represents the profit in dollars and x represents the number of hours worked. What is the girls' break-even point?

The students know to input the function in the [Y=] menu. They may need guidance in setting an appropriate window for the situation; then they graph the function, as shown in figure 27.1. The students also understand that the trace key can be used only to approximate the break-even point, which occurs where the function crosses the horizontal axis.

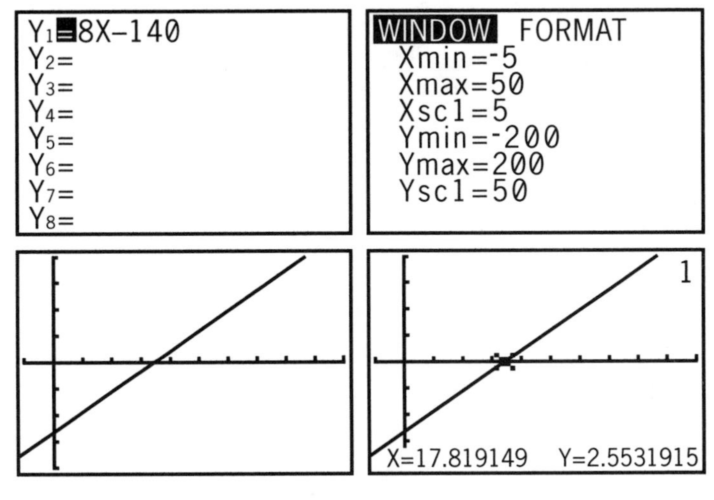

Fig. 27.1

Looking at the intersection point of the graph of the function and the horizontal axis, the students will know that the break-even point is between 17 and 18 hours. To find this point more accurately, the students can use the features of their table to find the x-value when $y = 0$, as shown in figure 27.2. By adjusting the step-value (ΔTbl), the students can zoom in to find the point (17.5, 0) on their table. Hence, after 17.5 hours of work, the girls will break even.

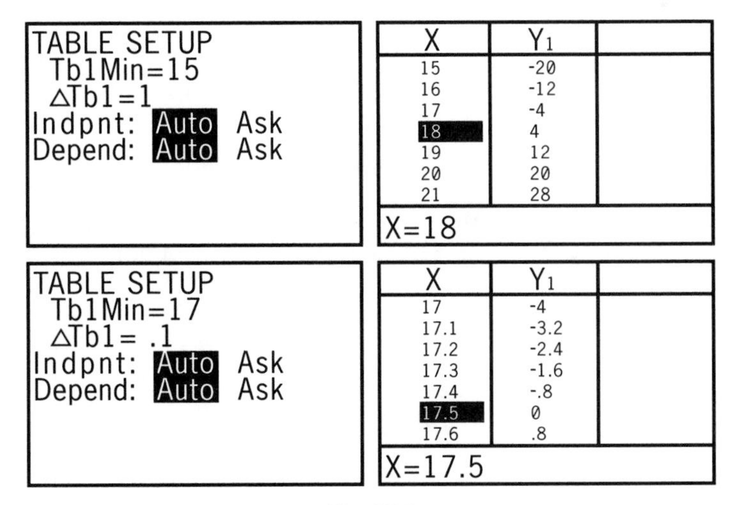

Fig. 27.2

To answer this break-even question by using Mathematica (1993), a student would enter the function rule as p[t_] = 8t – 140. The student is directed to think about the meaning of the input variable and to choose a reasonable domain for the situation. For example, the Plot[ ] command that is shown in figure 27.3 has a domain, chosen by the student, of 0 to 40 hours. Since this is an applications problem, we expect the students to label the input and output axes as part of their plot command. Notice that the students type the lines that are shown in bold Courier typeface. The computer output is shown immediately after each command.

```
p[t_]=8t -140
-140 + 8t

Plot[p[t],{t,0,40},AxesLabel->{hours,profit}]
```

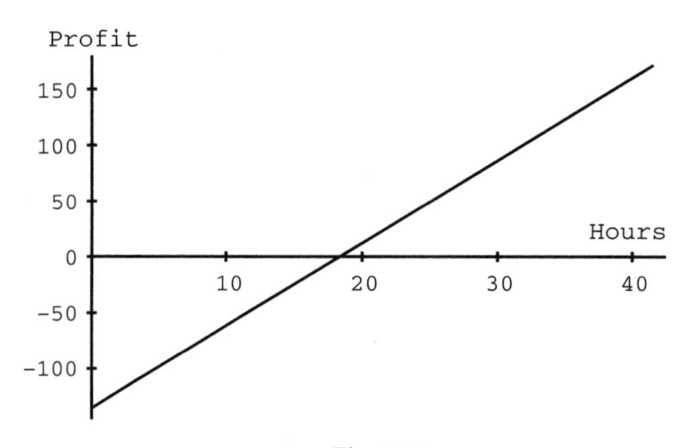

Fig. 27.3

Looking at this graph, our students would estimate that the break-even point occurs when t is about 18 hours. In using a graph for this purpose, they understand that their answer may be only an estimate and that they can improve their estimate by zooming in closer to the zero. For example, a student might change the domain to 15 to 20 hours, which produces the graph in figure 27.4. A better solution appears to be 17.5 hours. Later, the students also learn that they can solve the question of break-even points by finding t when $8t - 140 = 0$.

SOLVING AN INEQUALITY BY USING A TABLE OR A GRAPH

While studying quadratic functions, students are given problems like the following:

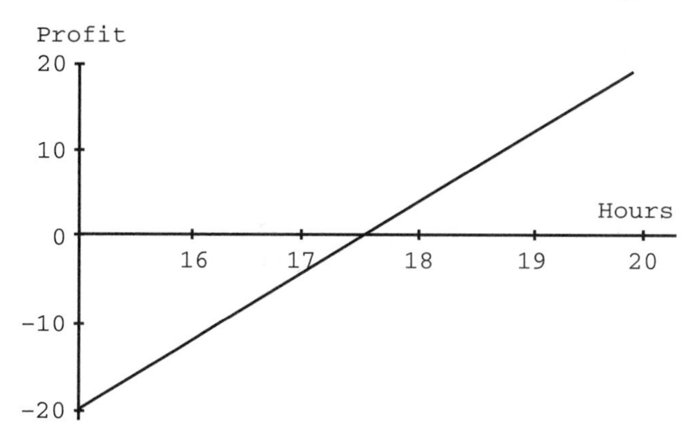

Fig. 27.4

Situation: Pop Fligh runs up and catches a ball. He then tosses it into the air from an initial height of 1.6 meters with an initial velocity of 10 meters per second. Recall that the height of an object in motion is a quadratic function of its time in the air. The rule for Pop's throw is $y = -4.9x^2 + 10x + 1.6$, where y represents the height of the ball in meters and x represents the time in seconds the ball has been in motion. When will Pop's fastball be *at least* 5 meters high?

For this situation, there are two equations to enter into the graphing calculator—Pop Fligh's function rule and the horizontal 5-meter line (see fig. 27.5). Some students may still need assistance in setting an appropriate window for this graph, but most have learned to set up their own windows by now, using data from the calculator's table as a guideline.

```
Y1=-4.9X²+10X+1.
6
Y2=5
Y3=
Y4=
Y5=
Y6=
Y7=
```

```
WINDOW FORMAT
  Xmin=-2
  Xmax=4
  Xscl=1
  Ymin=-10
  Ymax=10
  Yscl=5
```

Fig. 27.5

The students begin by examining the inequality $-4.9x^2 + 10x + 1.6 > 5$. They consider the two functions $y_1 = -4.9x^2 + 10x + 1.6$ and $y_2 = 5$. Two different algorithms can be used while working with most graphing calculators. First, the student can use the table feature to look for the independent variable that makes the first function and the second function both 5 meters. Since the first function is a quadratic function, there are two solutions, approximately 1.6 seconds and 0.45 second, as shown in figure 27.6. The students solve the problem by interpreting these results to mean that between 0.45 second and 1.6 seconds after Pop tossed the fly ball, it was at least 5 meters in the air.

X	Y_1	Y_2
1.4	5.996	5
1.5	5.575	5
1.6	5.056	5
1.7	4.439	5
1.8	3.724	5
1.9	2.911	5
2	2	5
X=1.6		

X	Y_1	Y_2
.3	4.159	5
.35	4.4998	5
.4	4.816	5
.45	5.1078	5
.5	5.375	5
.55	5.6178	5
.6	5.836	5
X=.45		

Fig. 27.6

The second strategy that can be used to answer the Pop Fligh question employs the graphing features of the calculator. The students can use the CALC menu to determine the two points where the first and second functions intersect. The graph provides the visual verification that the ball passed the 5-meter mark on the way up and on the way down (see fig. 27.7). The students recognize that the intersections of the parabola and the line will lead them to the solution to the problem. Once again, the students show their understanding of these algorithms by answering that approximately between 0.431 second and 1.610 seconds, the ball was at least 5 meters in the air.

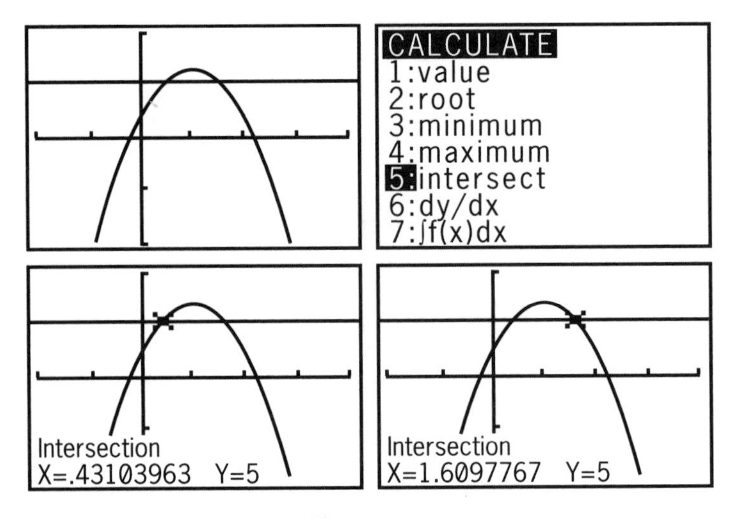

Fig. 27.7

Mathematica (1993) can use the same tabular and graphing algorithms as the graphing calculator. However, one additional strategy that Mathematica provides is the Solve[ ] command. After the students enter both functions, they can ask the program to find the point or points where the functions are equal to each other, as shown in figure 27.8.

```
Plot[{h1[t],h2[t]},{t,-2,4},AxesLabel->{time, height}]
```

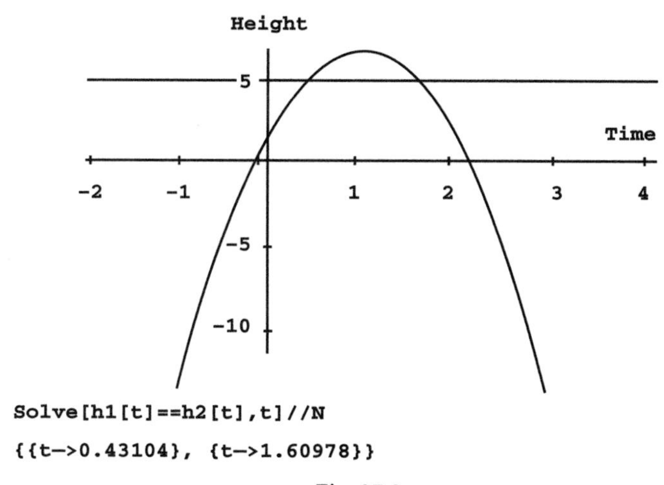


```
Solve[h1[t]==h2[t],t]//N
{{t->0.43104}, {t->1.60978}}
```

Fig. 27.8

Traditionally, the solution of quadratic and linear systems is a topic for advanced algebra. However, our introductory-algebra students showed that they could use the tabular and graphing algorithms provided by the technology to solve this advanced question with appropriate accuracy. What's more, our students might have a better concept of what the process and solution mean than a student who can merely move the symbols around to find an exact, but inappropriate, answer.

SOLVING A LINEAR EQUATION BY USING A TABLE

Mathematica tables provide a powerful algorithm for solving problems, either quadratic or linear. Recall the situation about the two girls and their lawn-mowing business. The students can answer other types of questions, such as "When will the girls make a profit of $64?" Figure 27.9 shows how a student can investigate the solution to this question with a table of ordered pairs. With the data presented this way, the students know that the first number in the pair represents the number of hours worked and the second number lists the total profit. When asked to find a profit of $64, the students would look for 64 as the second number in the pair. Since they are unable to find 64 as the profit, they must determine that 64 is between 60 and 68, both of which appear in the table they made. At this point the students can determine that the number of hours worked for a profit of $64 is somewhere between 25 and 26.

```
Table[{t,p[t]},{t,15,30,1}]
{{15,-20}, {16,-12}, {17,-4}, {18,4}, {19,12},
 {20,20}, {21,28}, {22,36}, {23,44}, {24, 52},
 {25,60}, {26,68}, {27,76}, {28,84}, {29,92}, {30,100}}
```

Fig. 27.9

Even though this problem is examined early on in our study of linear functions, some students know immediately that since this function is linear and since $64 is halfway between $60 and $68, the number of hours worked is 25.5. However, if students do not have this intuition, they can solve the problem by changing the table command. The students would start the table at a value of 25, stop it at 26, and have it proceed in steps of 0.1. With this approach, the students begin to zoom in on the output value desired so that they can find the input needed. The new table, shown in figure 27.10 in column format this time, provides the solution. Here the students can see that 25.5 hours worked produces a profit of $64. The students become proficient with this method of solving linear equations within the first two months of the school year.

```
Table[{t,p[t]},{t,25,26,0.1}]//TableForm
25       60
25.1     60.8
25.2     61.6
25.3     62.4
25.4     63.2
25.5     64
25.6     64.8
25.7     65.6
25.8     66.4
25.9     67.2
26.      68.
```

Fig. 27.10

SOLVING AN EQUATION WITH A SPREADSHEET

The students in our regular algebra course also used spreadsheets during their development of solutions of linear equations. One writing project in the middle of the second quarter required the students to compare the operating costs of an energy-efficient light bulb to those of a regular light bulb and find when the costs were the same (McConnell et al. 1996, p. 339).

Erin found light bulbs in a local store that were priced at \$0.75 for regular bulbs and \$1.00 for energy-efficient ones. From the information on the label and from her parents' electric bill, Erin figured that the cost of using the regular bulb was 0.48 cents per hour and the cost of using the energy-efficient bulb was 0.42 cents per hour. Following the recommendation in the textbook, she set up a spreadsheet showing 100-hour intervals. She wrote separate formulas for the cost column for each light bulb. The formula for the regular bulb was 0.75 + 0.0048∗Hours and that for the energy-efficient bulb was 1.00 + 0.0042∗Hours (the ∗ is the computer's symbol for multiplication). Erin's spreadsheet is shown in figure 27.11. She indicated where she thought that the two bulbs' total costs were the same. She also included a graph built from the spreadsheet. Erin reported that the cost of the bulbs was the same at 400 hours. Again, her answer is appropriate to the situation. Erin could have solved the system

$$y = 0.75 + 0.0048x$$
$$y = 1.00 + 0.0042x$$

by using her graphing calculator, or she could have used Mathematica's Solve [] command to find the exact solution to the equation 0.75 + 0.0048h =

Hours	Regular	Energy Efficient		
0	0.75	1.00		
100	1.23	1.42		
200	1.71	1.84		
300	2.19	2.26		
400	2.67	2.68	Cost of bulbs will be the same	
500	3.15	3.10		
600	3.63	3.52		
700	4.11	3.94		
800	4.59	4.36		
900	5.07	4.78		
1000	5.55	5.20		

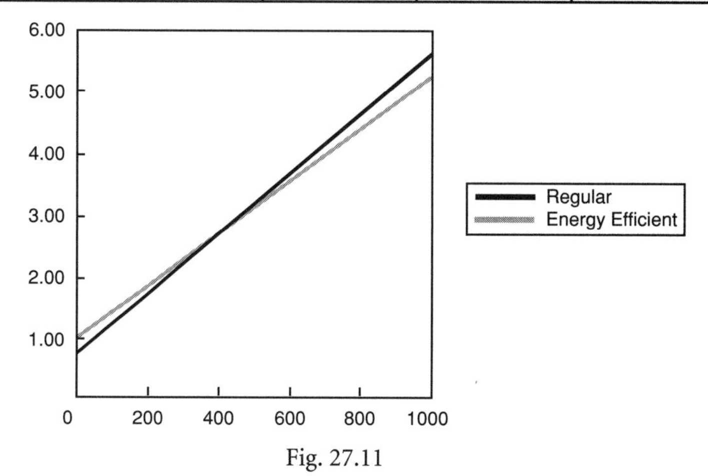

Fig. 27.11

1.00 + 0.0042h, but a more precise answer would add no useful information to her spreadsheet estimate of 400 hours. Further, the spreadsheet allowed her to answer more questions about the situation, such as "How much is saved with the energy-efficient bulb after 1000 hours?"

CONCLUSION

The new methods presented in this paper encourage an algorithmic approach to the instruction "to solve" and open the door to a wide variety of realistic questions about the problem situations. These questions lead to more applications and to richer ways of representing the problems. Teachers of introductory algebra can spend more time on the translation from a real-world problem to a symbolic equation. Algorithms based on computer and calculator technology prompt students to ask many questions about a situation that can be answered with reference to a table or a graph. Students then write about each solution and explain the meaning of their results. The students gain a deeper understanding of the role of algebra as a language that links technology to problems.

We conclude with a reference to Rugg and Clark's monograph (1918, p. 154) from eighty years ago:

> Our thesis is this: The central element in human thinking is seeing relationships clearly. In the same way the primary function of a secondary course in mathematics is to give ability to recognize verbally stated relationships between magnitudes, to represent such relationships economically by means of symbols, and to determine such relationships.

Rugg and Clark distinguish "tool outcomes" from "thinking outcomes." We believe that the general emphasis of this paper represents a thinking outcome that has great utility and can prepare students for the tool outcomes represented by the traditional interpretation of "to solve" in algebra.

REFERENCES

Fey, James, and M. Kathleen Heid. *Computer-Intensive Algebra.* College Park, Md.: University of Maryland, 1991.

Mathematica Ver. 2.2. Champaign, Ill.: Wolfram Research, 1993.

McConnell, John W., Susan Brown, Zalman Usiskin, Sharon L. Senk, Ted Widerski, Scott Anderson, Susan Eddins, Cathy Hynes Feldman, James Flanders, Margaret Hackworth, Daniel Hirschhorn, Lydia Polonsky, Leroy Sachs, and Ernest Woodward. *UCSMP Algebra.* 2nd ed. Glenview, Ill.: ScottForesman, 1996.

Rugg, Harold Ordway, and John Roscoe Clark. *Scientific Method in the Reconstruction of Ninth-Grade Mathematics.* Monograph of the *School Review,* vol. 2, no. 1. Chicago: University of Chicago Press, 1918.

28

A New Look at an Old Algorithm
The Semiaverage Line

Michael McNamara

THE term *algorithm* has, for many of us mathematics educators, a split personality. It often connotes the adjective *mechanical.* On the one hand, mechanical learning is something we have been trying to avoid for years. The terms *mechanical* and *algorithm* might cause us to think of the "problem solving" that went on when a teacher would state a problem for a class and then give the step-by-step procedure that all would use to solve the problem and then move on to the next problem. On the other hand, we want and need the mechanical nature of algorithms to save us time and mental effort. The subtraction algorithm I learned in first or second grade still serves my checkbook register and me very well (and too often). The laboratory technician who needs to make a solution from two stock solutions cannot afford the time to rethink the problem from the beginning; she needs, and probably has, an algorithm and does the mixing "mechanically."

We all know that we need algorithms, and we all want to avoid the pitfalls that come when thought and understanding intersect with the mechanical and the rote. I shall present a situation that involves fitting a line to several data points, a perfectly acceptable problem for students who have learned about lines and their equations.

Consider the data set in figure 28.1, giving the number of building permits issued in a small town over a ten-year period. After graphing the data, a class might make guesses about a line that seems to "fit." The class then might go home that night with a sheet that has the graph of the data and the graphs of two carefully selected lines (see fig. 28.2). The first part of the assignment could include the following tasks: (1) Determine which line fits the data better; (2) write a paragraph that makes the claim for the better line and explains your reasons in detail; and (3) write equations for each of the given lines. The second part of the assignment could require the students to invent a new set of data points that no line could possibly fit well.

Year	1	2	3	4	5	6	7	8	9	10
Number of permits	17	17	20	19	22	24	24	23	27	27

Fig. 28.1. A data set showing the number of building permits issued over a ten-year period

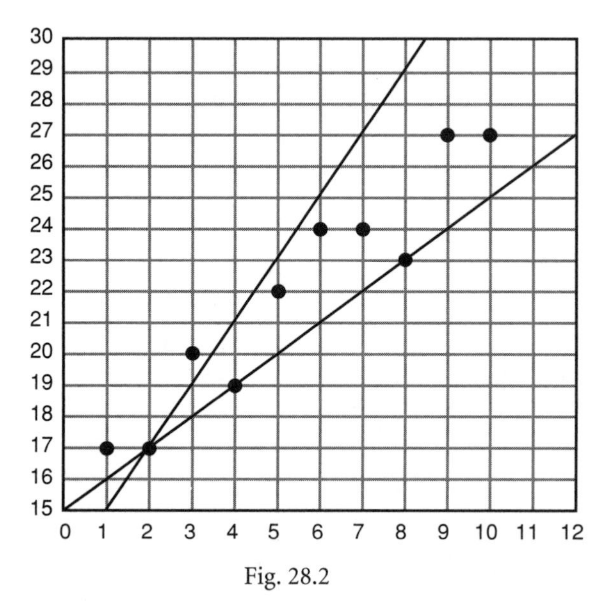

Fig. 28.2

The point is that many lines seem to fit well. Even if we do agree on the criterion for judging a good fit, we could perhaps never be sure that a given line *best* suits that criterion, and that is not our purpose here. With a dozen students fitting a line to the data, we might have a dozen good lines, all fitting well by some reasonable standard but with wildly different prediction values! One way out of the dilemma for students at this level of algebra is to agree to use a particular algorithm for a line of good fit: the "semiaverage line."

As its name implies, the algorithm calls for splitting the data set (which should first be ordered according to the sizes of its x-coordinates) into two halves (semi) and then determining the mean (average) for each coordinate in each half (see fig. 28.3). In the instance of an odd number of data points, the middle point is used in both halves. This semiaveraging then produces two points: one typical of the first part of the data and the other typical of the second part of the data. The semiaverage line is the line that contains these two points. For our sample data, the semiaverages give the points (3, 19) and (8, 25). The equation for the line that contains the points (3, 19) and (8, 25) is then found: $y = 1.2x + 15.4$. This exercise is an excellent application

First Half

Year	Number of permits
1	17
2	17
3	20
4	19
5	22

x-average 3   y-average 19

Second Half

Year	Number of permits
6	24
7	24
8	23
9	27
10	27

x-average 8   y-average 25

Fig. 28.3

of a recently learned algebraic skill. When the semiaverage points are plotted and the line is graphed (fig. 28.4), we can intuitively acknowledge it as a good fit. The equation for the line allows us to extrapolate, even beyond the frame of the graph.

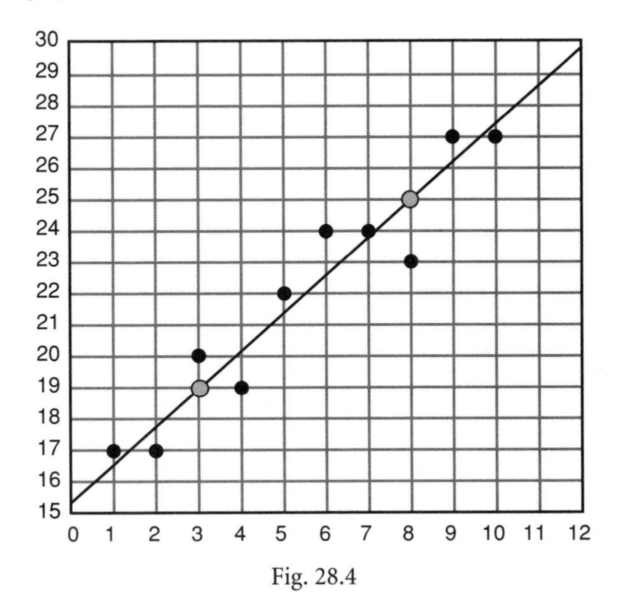

Fig. 28.4

The steps of the algorithm are easy to learn and fun to teach because students are learning a powerful and practical tool. Students might be encouraged to research their own data sets and make relevant predictions. What will be the cost of a Big Mac in five years? When will we be paying $20 for a movie ticket? Such topics as ticket prices, voter registrations, nursing home populations, library use, the number of teams in the National Hockey League, and mean SAT scores might become the objects of study and even develop into serious projects. In such investigations, algebra skills come alive—for some students for the first time—through a meaningful and reasonable algorithm.

BIBLIOGRAPHY

Mason, Robert D. *Statistical Techniques in Business and Economics.* Homewood, Ill.: Richard D. Irwin, 1974.

Emerson, John D., and David C. Hoaglin. "Resistant Lines for *y* versus *x*." In *Understanding Robust and Exploratory Data Analysis,* edited by David C. Hoaglin, Frederick Mosteller, and John W. Tukey, pp. 129–65. New York: John Wiley & Sons, 1983.

29

Random-Number Generators: A Mysterious Use of Algorithms

Stephanie O. Robinson

Donald J. Dessart

Most people think that randomness is an elusive concept that has very little to do with order! Students see algorithms, however, as orderly procedures or sequential steps that lead to precise answers. For example, using the division algorithm, we can show that 535 divided by 5 is precisely 107. How, then, is randomness related to algorithms? We will show that randomness is dependent on algorithms.

A Senior Class Problem and Randomness

The members of the senior class at Jefferson High School wish to choose 10 students from its class of 400 students to participate in a districtwide political event. All the seniors wish to participate. How can the 10 be selected so that each senior has an equal and fair chance of being chosen? John, a bright senior, had heard of random-number tables and suggested assigning each senior a whole number from 1 through 400 and then using the table of random numbers to pick the 10 participants.

How is a random-number table created and how is it related to algorithms? Webster (1993, p. 28) defines *algorithm* as "a procedure for solving a mathematical problem ... in a finite number of steps that frequently involves repetition of an operation." Thus, algorithms are involved in any mathematics problem where an answer is calculated by using a series of steps (Maurer 1991). Algorithms necessitate a systematic approach. The concept of random numbers and randomness seems to lie in direct contrast with the topic of algorithms. However, a closer look at randomness shows us that algorithms are at the very heart of the topic.

"Randomness surrounds us in everyday living. The results of a presidential poll, the toss of a coin, the winning numbers in a lottery, the spinning of a spinner are a few examples of common random phenomena" (Dessart 1995, p. 177).

It is the heart of probability and statistics, which are found in everyday life, in the social sciences, and in the physical and biological sciences. "Ideas of probability are fundamental to our understanding of the nature of life, of evolution and of the creation and behavior of matter" (Belsom et al. 1992, p. 7).

In this paper, we will attempt to answer the following questions:

1. What are randomness and random numbers?
2. What does randomness have to do with algorithms?
3. How can students generate random numbers?
4. How can students analyze a set of numbers for properties of randomness?

RANDOMNESS

What is randomness? Persi Diaconis, studying the physics of coin tossing, says that "the more you think about randomness, the less random things become" (Kolata 1986, p. 1068). Moore (1990, p. 98) gave this description: "Phenomena having uncertain individual outcomes but a regular pattern of outcomes in many repetitions are called *random*." Two necessary characteristics of randomness are (1) that the outcome cannot be predicted from a single occurrence of the phenomenon and (2) that a pattern emerges from many repetitions of the phenomenon. For example, when tossing a die with sides numbered from 1 through 6, we cannot predict whether the next throw will turn up a side with one dot. However, after tossing the die many times, we see that the one dot will appear about 100 times in 600 throws.

What is a random number? Single numbers are not random; only long sequences of numbers can be examined for randomness. A random number is defined to be "a series of digits where (*a*) each digit has the same probability of occurring, and (*b*) adjacent digits are completely independent of one another" (Landauer 1977, p. 296). Random numbers can be used to solve difficult problems by simulation. Although not particularly difficult, the problem involving the high school seniors is an example of the use of random-number tables to simulate the drawing of a name. Simulation depends on the fact that if an experiment is repeated a great number of times, the relative frequency of an observed phenomenon settles about a fixed value associated with that phenomenon (Dessart 1995). The tossing of a die is a standard technique for examining the concepts of probability. However, simulations take a long time and can become tedious. Imagine tossing a die 600 times to determine the number of times the one dot appears. The use of random-number tables makes this process much more time efficient.

In contrast with many mathematical models, randomness is an elusive subject. Many mathematical models are considered deterministic, meaning that they provide exact information that can be used to predict future information. By comparing the predictions of the model with the actual occurrences, we can assess the goodness of fit of the model, that is, how well the model predicts the reality. Statistical data do not provide exact information, so that one cannot predict future values with certainty but can use probability models that match predictable patterns.

CREATING RANDOM-NUMBER TABLES

How do we create random numbers? Since we are using an algorithm, we are not creating random numbers in the same way as one would in tossing a die. After the first number is chosen, the rest are determined by the algorithm, which is not so in tossing a die. We can call these numbers *pseudorandom* numbers. One problem with truly random numbers is that they can't be reproduced by the process that created them. In many instances, scientists need to be able to reproduce the exact random sequences used in simulations so that their work can be checked by others. Here are several algorithms devised and used to create pseudorandom-number tables.

The Middle-Square Algorithm

While working on the Manhattan Project in 1946, John von Neumann, Stanislaw Ulam, and N. Metropolis devised the middle-square algorithm, which simulates a roulette wheel, to create many random numbers (Myerson 1983). The steps to generate random numbers from 0000 to 9999 are the following:

1. Select any four-digit number, for example, 3214.
2. Square that number to obtain 10329796.
3. Select four digits from the middle of the number, that is, 3297.
4. The first digit, 3, is a single-digit random number, 32 is a two-digit random number, and so on.
5. Repeat steps 2, 3, and 4 with the new number, 3297.
6. Continue repeating the algorithm until the desired number of random numbers has been created (Meyerson 1983, p. 2).

The middle-square algorithm was acceptable for its time. However, a major problem is that the numbers unavoidably repeat themselves. The sequence of such numbers is called a *cycle,* or *period.* Once a cycle begins, then the numbers following it will repeat. Sooner or later numbers will repeat themselves, since there are only a finite number of possibilities.

THE LINEAR CONGRUENTIAL ALGORITHM

In 1948, D. H. Lehmer devised another method, called the *linear congruential generator,* which reduced the frequency of cycling (Meyerson 1983). Below is the algorithm with an example:

1. Multiply a starting seed number by a constant. For example, multiply 3214 by 4, which gives 12856.
2. Add another constant, for example, 78, which gives 12934.
3. Reduce the result by a modulus m (divide the result by m and retain only the remainder). In our example using a modulus of 43, divide 12934 by 43, leaving a remainder of 34.

Our new number is 34. Now we use 34 as a seed and repeat steps 1, 2, and 3. We get 42 for the next seed. Now we have 34 42 as the beginning of our random-number table. We should obtain numbers between 0 and $(m - 1)$, one less than the modulus. Cycling occurs with at most m iterations, with m being the number chosen for the modulus. In our example we would get numbers from 0 to 42, with cycling occurring within forty-three iterations. The modulus limits the maximum range of the numbers. However, the choice of constants for steps 1 and 2 determines the number of iterations before cycling occurs. By choosing a large modulus and specific constants to multiply and add, we can get the maximum of m nonrepeating numbers. One popular choice of constants for the algorithm is 16807 for the multiplier in step 1, 0 for the additive constant in step 2, and for the modulus, $2^{31} - 1$, the greatest prime number that can be represented in a thirty-two-bit computer (Deng and Rousseau 1991).

Some new random-number generators produce tables with extremely long periods. For example, suppose that a computer can generate 1 000 000 random numbers per second, using a new type of generator called a *matrix congruential generator.* The computer needs more than 140 000 years to complete the cycle, compared to just 36 minutes for a cycle with Lehmer's congruential generator (Deng and Rousseau 1991, p. 92).

Shift-Register Algorithms

Other techniques, known since the 1950s, are called *shift-register algorithms.* In this type of algorithm, each newly generated number depends on several previous numbers in the list and a chosen modulus m. A simplified example of this method is based on a Fibonacci sequence. In a Fibonacci sequence, the two previous numbers are added to create the next, that is, $X_n = X_{n-1} + X_{n-2}$. The sequence begins with two 1s: 1, 1, 2, 3, 5, 8, 13, 21, 34, 55, 89, ... In our example of a shift-register algorithm, we add the two previous numbers in modulus 7, which generates the sequence 1, 1, 2, 3, 5, 1, 6, 0, 6, 6, 5, ... The addends, the numbers to be added, can also be separated by one or more numbers in the list. Another sequence in modulus 7 might begin with three 1s, with $X_n = X_{n-1} + X_{n-3}$. In this example the sequence would be 1, 1, 1, 2, 3, 4, 6, 2, 6, 5, ... Since the numbers from which the new entry is formed change each time by adding the new number and discarding another, the algorithm is called a *shift register.*

Although relatively simple to build, a series of numbers based on a Fibonacci type of algorithm does not test well for randomness. Testing for randomness will be discussed later in the paper. Other shift-register algorithms, using different moduli and choices of addends, result in random-number tables that do test well for randomness. The selection of the initial numbers in the sequence and the modulus are essential to building a random-number table that tests well.

The 147 Algorithm

Students can use the 147 algorithm with a computer or simple handheld calculator. The steps are as follows:

1. Select any seven-digit number preceded by a decimal point and with the last digit odd and other than 5 (e.g., 0.2415967).
2. Multiply the decimal obtained in step 1 by 147 (which in this example gives 35.5147149).
3. From that product, subtract the whole-number portion of the decimal (i.e., 35.5147149 – 35 = 0.5147149). Thus 5 is the first random number, 51 is a two-digit random number, and so on.
4. Continue this process by repeating steps 2 and 3 to obtain a sequence of random numbers (Råde 1981, p. 119).

See figure 29.1 for the steps to program the TI-81 or TI-82 calculator to generate this algorithm.

We have used an algorithmic approach to create series of random (pseudorandom) numbers. One of the advantages of using an algorithmic approach is that we can analyze and compare the efficiencies of many algorithms that solve the same problem. Students can compare the algorithms for efficiency by using various tests for randomness, as described in the next section.

```
Program Instructions
:Disp "X="
:Input X
:Lbl 1
:147*X→N
:N-iPart N→N
:Disp N
:N→X
:Pause
:Goto 1
```

Fig. 29.1. Random-number-generator program for the 147 algorithm for use with a TI-81 or TI-82 calculator

TESTING FOR RANDOMNESS

We have explored some algorithms that generate sequences of numbers. We claim that these numbers are random. But are they truly random? Unfortunately, there are no methods that "prove" that a set of numbers is, in fact, random. We can only "test" for randomness. Many tests have been developed, and no sequence has been known to satisfy all these tests (Kolata 1986)!

Tests for randomness are based on what are considered properties of random numbers. The first characteristic is uniform distribution, which means that the digits have equal representation within the sequence. Uniform distribution can be checked by a frequency test. For single digits, each digit occurs about 10 percent of the time. You could also partition the sequence into pairs or triplets and check the distributions (Seife 1997). For pairs from 00 to 99, each pair would occur an equal number of times, or 1 percent for each pair.

The second major characteristic to examine when testing for randomness is independence of digits, that is, that the next digit cannot be predicted on the basis of the previous digit. One test for independence is the up-down, or high-low, test, comparing a number with the one following it. How many times is the next number greater? How many times is the next number less? Random numbers have these characteristics: they remain the same about 10 percent of the time, increase 45 percent, and decrease 45 percent (Hoffman 1979).

Another test for randomness is called the *poker test,* in which the sequence is partitioned into sets of four digits each. Within each four-digit set, we determine whether any of the following exist: one pair, two pairs, three of a kind, four of a kind, or no repeated digits. If the numbers are random, the occurrence will approximate the probabilities in table 29.1 (Råde 1981, p. 122).

TABLE 29.1
Probabilities of Occurrence of Types of Four-Digit Sets

Types of Four-Digit Sets	Example	Probability
No repeated digits	2346	0.504
One pair	2344	0.432
Two pairs	3355	0.027
Three of a kind	3337	0.036
Four of a kind	4444	0.001

An Example Using the 147 Algorithm

The 147 method was used to obtain a series of 100 digits (see fig. 29.2). This series can be examined for uniform distribution of digits, pairs, or triplets. The high-low test for independence can also be applied, as well as the poker test (see tables 29.2, 29.3, and 29.4 for the results of these tests). You can generate your own 100-digit table and compare your results with these. But don't be too disappointed with your results! Experts such as Diaconis think that in all likelihood, new methods for generating random numbers will fail some tests for randomness (Kolata 1986). We can only

51 47 15 66 31 05 47 64 35 03
59 45 28 39 15 73 55 05 11 92
35 52 75 45 54 91 15 71 99 05
82 60 85 42 71 45 79 03 15 17
63 05 91 68 35 77 47 45 88 75

Fig. 29.2. A 100-digit random-number table generated by the 147 algorithm

Table 29.2
The Distribution of the 100 Digits in Figure 29.2

Digit	Frequency of Occurrence
0	7
1	13
2	5
3	9
4	10
5	24
6	6
7	12
8	6
9	8

Table 29.3
The Up-Down Test for the Independence of Digits in Figure 29.2

Occurrence	Frequency
The digit remained the same.	9
The digit was greater.	46
The digit was less.	44

Table 29.4
Poker Test for Digits in Figure 29.2

Case	Frequency	Probability
All digits different	11	0.44
One Pair	13	0.52
Two Pair	0	0
Three of a Kind	1	0.04
Four of a Kind	0	0

hope to find a sequence that will pass many of the tests for randomness, such as those described above.

IMPLICATIONS FOR INSTRUCTION

The topic of random numbers and the algorithms to create and test for randomness provide a rich avenue for activities with algorithms. Students can use generators to create random-number tables, create their own algorithms to generate random numbers, and analyze the efficiency of the various methods by testing their results for the characteristics of randomness. A discussion of comparisons of student-explored algorithms to create random-number tables will enhance students' opportunities to communicate mathematically and to make decisions based on logical reasoning.

Connections can be made to a wide variety of topics, including simulation, probability, statistics, social sciences, and physical and biological sciences. Students can compare algorithms in hands-on activities using spinners, dice, coins, and so forth, and thus develop intuitive notions of randomness and an appreciation of statistical methods.

Random-number generation is a topic that sparks the interest of students in many ways. It allows students to apply algorithms, analyze results, and compare theoretical and experimental activities. By using the methods of creating random numbers and analyzing lists of numbers for randomness, students engage in problem solving, communicating, connecting, and reasoning. And they should have fun doing it!

REFERENCES

Belsom, Chris, Stan Dolan, Chris Little, and Mary Rouncefield. *Living with Uncertainty.* Cambridge: Cambridge University Press, 1992.

Deng, Lih-Yuan, and Cecil Rousseau. "Recent Development in Random Number Generation." In *Proceedings of the Twenty-ninth ACM Annual Southeast Regional Conference,* edited by William Day, Kai Chang, and Carolyn McCreary, pp. 89–94. New York: ACM Press, 1991.

Dessart, Donald J. "Randomness: A Connection to Reality." In *Connecting Mathematics across the Curriculum,* 1995 Yearbook of the National Council of Teachers of Mathematics, edited by Peggy A. House, pp. 177–81. Reston, Va.: National Council of Teachers of Mathematics, 1995.

Hoffman, Dale T. *Monte Carlo: The Use of Random Digits to Simulate Experiments.* Lexington, Mass.: Consortium for Mathematics and Its Applications, 1979.

Kolata, Gina. "What Does It Mean to Be Random?" *Science,* 7 March 1986, pp. 1068–70.

Landauer, Edwin. "Methods of Random Number Generation." *Two-Year College Mathematics Journal* 8 (November 1977): 296–303.

Maurer, Stephen B., and Anthony Ralston. "Algorithms: You Cannot Do Discrete Mathematics without Them." In *Discrete Mathematics across the Curriculum, K–12,* 1991 Yearbook of the National Council of Teachers of Mathematics, edited by Margaret J. Kenney, pp. 195–206. Reston, Va.: National Council of Teachers of Mathematics, 1991.

Meyerson, Mark D. *Random Numbers.* Lexington, Mass.: Consortium for Mathematics and Its Applications, 1983.

Miriam-Webster's Collegiate Dictionary, 10th ed., s.v. "algorithm."

Moore, David S. "Uncertainty." In *On the Shoulders of Giants: New Approaches to Numeracy,* edited by Lynn Arthur Steen, pp. 95–137. Washington, D.C.: National Academy Press, 1990.

Råde, Lennart. "Random Digits and the Programmable Calculator." In *Teaching Statistics and Probability,* 1981 Yearbook of the National Council of Teachers of Mathematics, edited by Albert P. Shulte, pp. 118–25. Reston, Va.: National Council of Teachers of Mathematics, 1981.

Seife, Charles. "New Test Sizes Up Randomness." *Science,* 25 April 1997, p. 532.

30

Algorithmic Problem Solving in Discrete Mathematics

Eric W. Hart

DISCRETE mathematics is an important branch of mathematics that has been widely recommended for inclusion in the school curriculum (National Council of Teachers of Mathematics [NCTM] 1989; Dossey 1990; Hart 1991; NCTM 1991; Rosenstein 1997). Algorithms and algorithmic problem solving play a fundamental role in discrete mathematics. This paper will discuss algorithmic problem solving and illustrate how it is developed in the discrete mathematics strand of the high school curriculum from the Core-Plus Mathematics Project, a curriculum-development project funded by the National Science Foundation (NSF grant no. MDR-9255257).

Some important areas of discrete mathematics are (*a*) graph theory—using vertex-edge diagrams to study relationships among a finite number of elements, as in a transportation network or a predator-prey food web; (*b*) game theory—the mathematics of, for example, voting, fair division, apportionment, and cooperation and competition; (*c*) combinatorics—systematic counting; and (*d*) recursion—the method of describing sequential change by indicating how the next stage of a process is determined by previous stages. Other important discrete mathematics topics are matrices and discrete probability. Central themes in all contexts and areas are *existence* (does a solution exist?), *algorithmic problem solving* (can a solution be efficiently constructed?), and *optimization* (which is the best?).

This paper will examine the theme of algorithmic problem solving. Algorithmic problem solving is the process of designing, using, and analyzing algorithms for solving problems. Problem solving is an important skill that all students of mathematics should acquire. Algorithmic problem solving is a particular style of problem solving that can uniquely contribute to students' mathematical power. The examples below show how the skill of algorithmic problem solving can be developed through students' investigation of applications involving graph theory and recursion.

ALGORITHMIC PROBLEM SOLVING IN GRAPH THEORY

A *graph* in graph theory is a diagram consisting of *vertices* and *edges* between some of the vertices. In almost all graph-theory problems it is necessary to use algorithmic problem solving to reach a solution. This process is illustrated below by two fundamental problems in graph theory—finding an Euler circuit and finding a minimal spanning tree. In each example, an illustrative situation is presented, followed by a discussion of the algorithms involved, some sample student activities, and a follow-up discussion reflecting on the development and use of algorithmic problem solving.

Euler Circuits

Sample situation

Suppose you have the job of painting all the lockers on the first floor of a high school (see the floor plan in figure 30.1). The lockers are located along the walls of the halls. Five-gallon buckets of paint, a spray-paint compressor, and other equipment are located in the first-floor equipment room. You must move this bulky equipment with you as you paint the lockers. You must also return it to the equipment room when you have finished painting. If the lockers are on both sides of a hallway, they must be painted one row at a time so that the entire hallway is not blocked by the painting equipment. Since you are being paid by the job, not by the hour, you would like to paint the lockers as quickly and efficiently as possible. How can you do this?

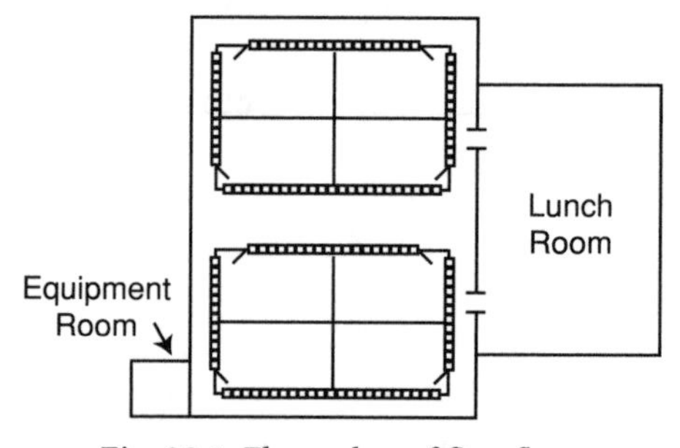

Fig. 30.1. Floor plan of first floor

Students can use vertex-edge graphs and algorithmic problem solving to solve this problem. The following instructional sequence can be used to help students find a solution: first, students "play around" with the problem to get a feel for what is involved; next, they build a mathematical model, which in this instance will be a vertex-edge graph; finally, they figure out how to use the model to get a solution.

The goal of the "playing around" step in this instructional sequence is for students to generate a list of criteria for an optimal locker-painting plan. For example, every row of lockers must get painted, the painter must start and end at the equipment room, and the painting equipment should not be moved past a row of lockers that are not being painted.

In the next step, students build a graph model that represents the given arrangement of lockers. A crucial part of building any graph model is to decide what the vertices and edges should represent. Students might let the vertices represent the points where the painter would stop painting one row of lockers and start painting another, and the edges could represent the rows of lockers. Figure 30.2 shows a graph model that represents the locker-painting problem for the first-floor lockers in this way. The final step in the solution process is to use the graph model to solve the problem. This is where Euler circuits and algorithmic problem solving come in.

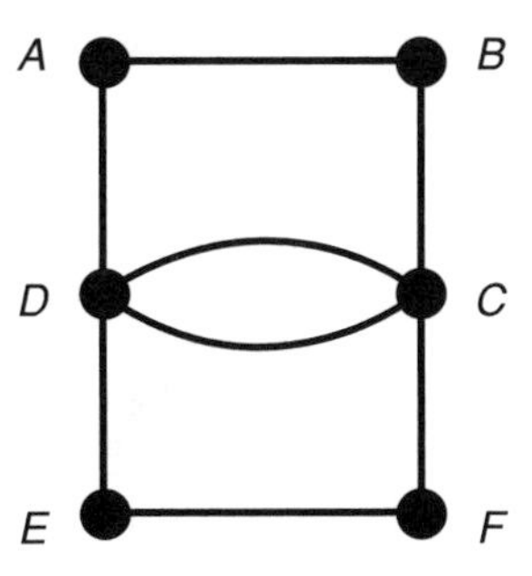

Fig. 30.2. Graph model of first-floor locker-painting problem

The algorithms

The criteria for the optimal locker-painting plan are the defining characteristics of an Euler circuit. An *Euler circuit* is a route through a connected graph such that (1) each edge of the graph is traced exactly once and (2) the route starts and ends at the same vertex. Any route that starts and ends at the same vertex without repeating edges is a circuit, but it may not trace all paths. Should it do so, it would be an Euler circuit. To solve the locker-painting problem, students must find an Euler circuit, which raises several important questions. Does every graph have an Euler circuit? (No) If not, how can you tell if a graph has an Euler circuit? (A connected graph has an Euler circuit if and only if the number of edges touching each vertex is even.) For a graph that has an Euler circuit, how can we find the circuit efficiently? To answer the last question we need a systematic procedure, that is, an algorithm.

At least two "standard" algorithms can be used to construct an Euler circuit. Hierholzer's algorithm is sometimes called the *onion-skin algorithm* because it works by peeling away shorter circuits; then all the shorter circuits are put together to form an Euler circuit. This is how it works: Start anywhere and "follow your nose" until you end up back where you started without retracing any edges. If you have used all the edges, you are done. If you have not used all the edges, choose any vertex on your initial circuit that is incident to an untraced edge. Start at this vertex and follow your nose until you get back to the vertex. Continue building shorter circuits in this way until all edges have been traced. Then link all the shorter circuits to get an Euler circuit. This algorithm is illustrated in student activity 3 below.

A very elegant, but less intuitive, algorithm is called *Fleury's algorithm.* Start at any vertex. Find an edge to trace so that all the untraced edges form a connected graph. Repeat until you have traced all the edges.

Euler-circuit problems provide a natural and interesting way for students to develop the skill of algorithmic problem solving, as seen in the activities below. Furthermore, the problems require little in the way of prerequisite knowledge, so all students can get productively engaged.

Some student activities

In the following activities, students begin to explore the properties of Euler circuits and the algorithms for finding them. This sequence of activities would typically be done by teams of students working together in small groups. The activities begin after students have had a chance to investigate Euler circuits in the locker-painting problem and in other contexts, such as finding an optimal route for collecting money from parking meters on downtown streets.

1. By looking at the form of an equation, you can often predict the shape of the graph of the equation without plotting any points. Similarly, it would be helpful to be able to examine a vertex-edge graph and predict if it has an Euler circuit without trying to trace it.
   (*a*) Each member of your group should draw a graph with five or more edges that has an Euler circuit. On a separate sheet of paper, draw a graph with five or more edges that does *not* have an Euler circuit.
   (*b*) Sort your group's graphs into two piles—those that have an Euler circuit and those that do not.
   (*c*) Carefully examine the graphs in the two piles. Describe the important ways in which the graphs with Euler circuits differ from those with no Euler circuit.
   (*d*) Try to figure out a way to predict if a graph has an Euler circuit simply by examining its vertices. Check your method of prediction by using graphs from previous activities.
   (*e*) Make a conjecture about the properties of a graph that has an Euler circuit. Explain why you think your conjecture is true for *any* graph with an Euler circuit.
2. Once you can predict if a graph has an Euler circuit, it is often still necessary to find the circuit. Consider the graphs in figure 30.3.

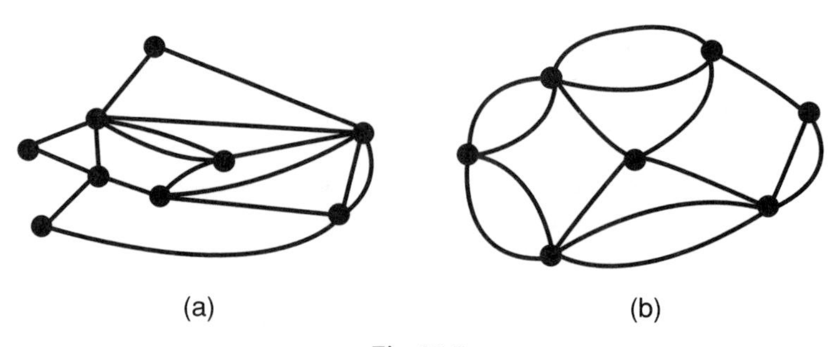

Fig. 30.3

(*a*) For each graph, predict whether it has an Euler circuit. (Answer: The graph in fig. 30.3a has all even vertices, so it has an Euler circuit.The graph in fig. 30.3b has some odd vertices, so it does not have an Euler circuit.)

(*b*) Find an Euler circuit in the graph that has one.

(*c*) Describe the method you used to find your Euler circuit. Describe other possible methods you can think of for finding Euler circuits.

3. One systematic method for finding an Euler circuit is to trace the circuit in several stages. For example, suppose you and your classmates want to find an Euler circuit that begins and ends at *A* in the graph on the left below. You can trace the circuit in three stages.

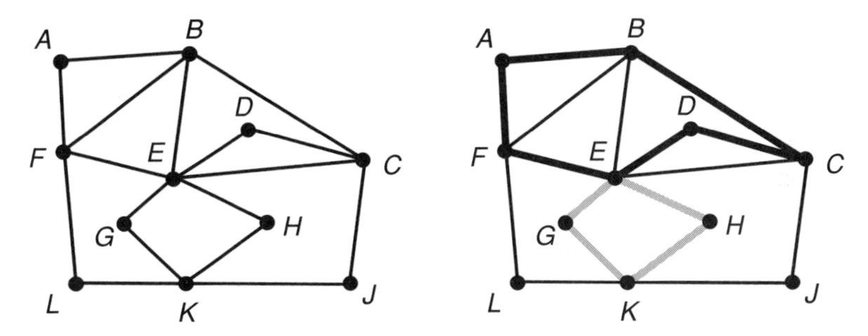

Stage 1: Alicia began by drawing a circuit that begins and ends at *A*. The circuit she drew, shown in the diagram by the heavy edges, was *A-B-C-D-E-F-A*. But this circuit does not trace all edges.

Stage 2: George added another circuit shown by the gray edges starting at *E: E-G-K-H-E.*

(*a*) Alicia's and George's circuits can be combined to form a single circuit beginning and ending at A. List the order of vertices for that combined circuit.

Stage 3: Since this circuit still does not trace each edge, a third stage is required.

(*b*) Trace a third circuit that covers the rest of the edges.

(c) Combine all the circuits to form an Euler circuit that begins and ends at *A*. List the vertices of your Euler circuit in order.

4. Choose your preferred method for finding Euler circuits.

(*a*) Write specific step-by-step instructions that describe the method you chose. Your instructions should be written so they apply to *any* graph, not just the particular one that you may be working on at the moment. Such a list of step-by-step instructions is called an *algorithm.*

(*b*) Compare your algorithm to those of other students or groups. Do they all seem to work? Do you think some are better than others? Explain.

Reflecting on the student activities

In these sample activities, students investigate algorithms for constructing Euler circuits. That is, they are engaged in algorithmic problem solving. Notice that algorithmic problem solving does not mean that students memorize a given algorithm and then apply it in many practice problems. Rather, students attempt to solve a problem by designing, using, and analyzing systematic solution procedures. In this sequence of activities, students first work on their own algorithms, then they are presented with a partially worked-out standard algorithm (Hierholzer's algorithm). Thus, they build the skills of constructing their own algorithms and working with given algorithms. Furthermore, they begin to analyze and compare algorithms. All of this is part of the developing skill of algorithmic problem solving. Another example follows.

Minimal Spanning Trees

Sample situation

In this information age, it is essential to find the best way to stay informed. You need to have the right information at the right time in order to make the best decisions and take the most effective action. One way to keep informed is through computer networks. Computers are no longer isolated in separate, unconnected locations; they are linked together in networks so that information can be shared among many users. In fact, there is a common saying that "the network *is* the computer."

Suppose that at a local high school six computers in six different offices are to be networked. The staff decide to link all the computers to one another without any kind of separate junction box or server. They would like to use the least amount of wire to link all the computers. Since electronic signals move so quickly, the connections between two computers are equally efficient whether they are linked directly or indirectly.

The matrix at the right shows which computers can be linked directly as well as how much wire is needed. Because of the location of the offices and the computers, it is not possible to run wire directly between every pair of computers. The computers are represented by letters, and the distances are in meters. What is the minimum amount of wire needed to connect all six computers so that every computer is linked directly or indirectly to every other computer?

	A	*B*	*C*	*D*	*E*	*F*
A	—	9	—	—	—	3
B	9	—	8	—	8	11
C	—	8	—	3	5	—
D	—	—	3	—	6	11
E	—	8	5	6	—	9
F	3	11	—	11	9	—

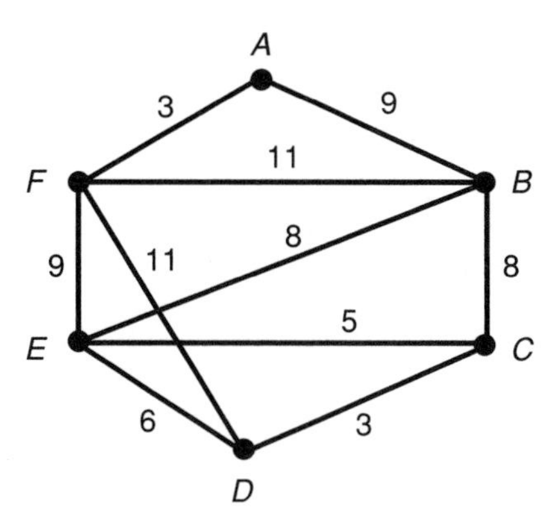

Fig. 30.4. A graph model of the computer-network problem

This is another problem that can be solved with vertex-edge graphs and algorithmic problem solving. Figure 30.4 shows a graph model of this situation. This graph is called a *weighted graph* because there are "weights" on the edges. In this instance, the vertices represent the computers, two vertices are connected by an edge if the two computers are directly connected, and the weight on an edge is the length of the wire connecting the two computers. To find a solution to the computer-network problem, we need to find a *minimal spanning tree* in the graph.

The algorithms

A minimal spanning tree in a connected graph is, roughly, a network within the graph that joins all the vertices and has minimum total weight. More precisely, a *connected* graph is a graph that is all in one piece. A *tree* is a connected graph that contains no circuits. A *spanning tree* in a connected graph G is a subgraph of G that includes all the vertices of G and is also a tree (see fig. 30. 5). A *minimal spanning tree* in a connected graph G is a spanning tree in G of minimum total weight.

There are several known algorithms for efficiently finding a minimal spanning tree. There are also some plausible algorithms that do not work. Three algorithms are considered here. Each of the three algorithms has the same basic strategy: successively add edges of minimum weight in such a way that no circuits are created. Slightly different elaborations of this strategy yield two algorithms that work and one that does not.

In Kruskal's algorithm, the edge added at a given step does *not* have to be connected to any previously added edges. In Prim's algorithm, the edge added at a

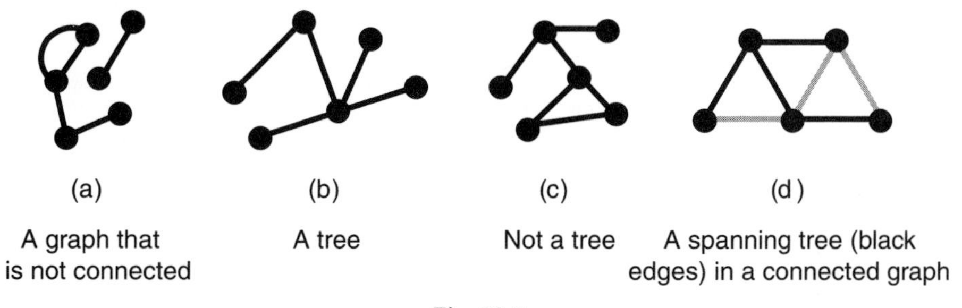

Fig. 30.5

given step must be connected to *some* previously added edge. In the nearest-neighbor algorithm, the edge added at a given step must be connected to the edge added in the immediately preceding step. Kruskal's and Prim's algorithms will efficiently produce a minimal spanning tree in any connected graph, but the nearest-neighbor algorithm may fail to produce a minimal spanning tree.

A more detailed statement of Kruskal's algorithm can be found in student activity 5 below, where it is referred to as the *best-edge algorithm*. See student activity 7 for a more precise statement of the nearest-neighbor algorithm. The steps of Prim's algorithm are as follows. Step 1: Choose any vertex to be the starting vertex. Step 2: Add an edge of minimum weight that connects to any vertex already reached, as long as it does not create a circuit. Step 3: Repeat step 2 until all vertices have been reached.

Students working on minimal-spanning-tree problems will often come up with all three of these algorithms, and the discussion about which work and which do not can be very rich, as illustrated in the student activities that follow.

Some student activities

This series of activities refers to the computer-network problem. Students begin by considering the computer graph in figure 30.4. Once again, it is generally advantageous to have students work in small groups.

1. Working in pairs within your group, solve the computer-network problem as follows:

   (*a*) What is the minimum amount of wire needed to connect the computers so that every computer is linked directly or indirectly to every other computer? (The students solve the problem any way they like.)

   (*b*) Make a copy of the computer graph and darken the edges of a shortest network.

   (*c*) Write a description of a method for finding your shortest network.

   (*d*) Compare your shortest network and the minimum amount of wire needed to what the other members of your group found. Discuss and resolve any differences. (It is likely that different students will generate different shortest networks. However, they should all have the same length.)

2. In activity 1*c*, you wrote a description of a method for finding a shortest network.

   (*a*) Exchange written descriptions with another pair of members in your group. Make a new copy of the graph and try to use their method to find a shortest network.

   (*b*) Does each method work?

   (*c*) Work together as a group to refine the methods, and then write down a step-by-step procedure—an *algorithm*—for each method that works.

3. Think about the properties of the shortest wiring networks you have been investigating. State whether each of the following statements is true or false. For each statement, justify your answer. Compare your answers to those of other groups and resolve any differences.

   (*a*) There is only one correct answer possible for the minimum amount of wire needed to connect all six computers. (T)

   (*b*) There can be more than one shortest network for a given situation. (T)

   (*c*) There is more than one algorithm for finding a shortest network. (T)

   (*d*) A shortest network must be all in one piece. (Graphs that are all in one piece are said to be *connected.*) (T)

   (*e*) All vertices must be joined by the network. (T)

   (*f*) A shortest network cannot contain any circuits. (A *circuit* is a path that starts and ends at the same vertex and does not repeat any edges.) (T)

4. A connected graph that has no circuits is called a *tree.*

   (*a*) Why does it make sense to call such a graph a tree?

   (*b*) A *minimal spanning tree* in a connected graph is a tree that has minimum total length and spans the graph; that is, it includes every vertex. Explain why the shortest networks you have found in the computer graph are minimal spanning trees.

5. There are, as you may have concluded in activity 3, several possible algorithms for finding a minimal spanning tree in a connected graph. Study the one below:

      (i) Draw all the vertices but no edges.

      (ii) Add the shortest edge that will not create a circuit. If there is more than one such edge, choose any one. The edge you add does not have to be connected to previously added edges, and you may to use more than one edge of the same length.

      (iii) Repeat step ii until it is no longer possible to add an edge without creating a circuit.

   (*a*) Follow the steps of this algorithm to construct a minimal spanning tree for the computer graph.

   (*b*) Explain why this algorithm could be called the *best-edge algorithm.*

   (*c*) Compare the minimal spanning tree you get with this best-edge algorithm to the one you found in activity 1. How do the lengths of the minimal spanning trees compare?

6. Examine the graph to the right.

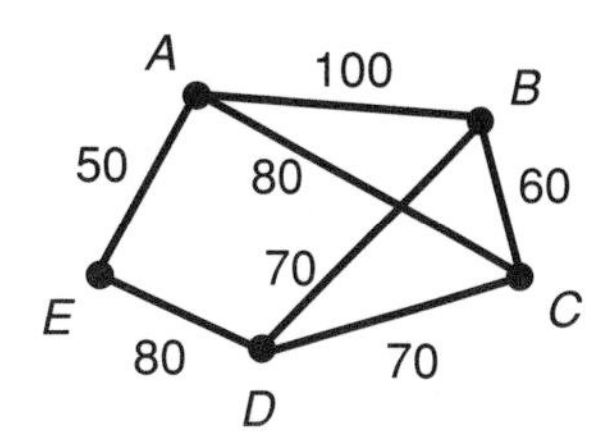

(*a*) Use the best-edge algorithm to find a minimal spanning tree for this graph. Calculate its length. (260)

(*b*) Explain why the best-edge algorithm can produce different minimal spanning trees. (Some steps pose a choice of different edges with the same length.)

(*c*) Find all possible minimal spanning trees. Compare their lengths. (260)

(*d*) How is the best-edge algorithm similar to, or different from, the algorithms you produced in activity 2?

7. Students in one class claimed that the following algorithm will produce a minimal spanning tree in a given graph.

   (i) Make a copy of the graph with the edges drawn in lightly.

   (ii) Choose a starting vertex.

   (iii) From the vertex where you are, darken the shortest edge that will not create a circuit. (If there is more than one such edge, choose any one.) Then move to the end vertex of that edge.

   (iv) Repeat step iii until all vertices have been reached.

   To test this algorithm, complete the following steps *a–f*. First, make four copies of the graph at the right.

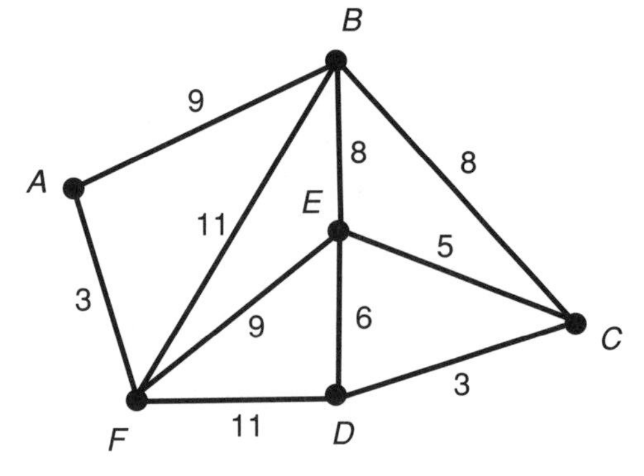

   (*a*) Apply the algorithm to the graph starting with vertex *E*. What is the total length of the network you get? (31)

   (*b*) Explain why this algorithm could be called the *nearest-neighbor algorithm.*

   (*c*) Apply the algorithm starting with vertex *C*. Record the length of the resulting network. (29)

   (*d*) Apply the algorithm starting with vertex *A*. What happens? (Proceeding beyond *A-F-E-C-D* creates a circuit.)

   (*e*) Now use the best-edge algorithm described in activity 5 to find a minimal spanning tree for this graph. (The length is 28.)

   (*f*) Do you think the nearest-neighbor algorithm is a good method for finding a minimal spanning tree? Write a brief justification of your answer.

8. How are the nearest-neighbor algorithm and the best-edge algorithm similar? How are they different?

9. Two important questions about any algorithm are, Does it always work? and Is it efficient? You will continue to investigate these questions as you study more about graphs. For the best-edge algorithm, mathematicians have proved that the answer to both questions is yes. That is, the best-edge algorithm will efficiently find a minimal spanning tree for any connected graph. What are your thoughts about these questions for the nearest-neighbor algorithm?

   (*a*) Does it always work?

   (*b*) Is it efficient?

Reflecting on the student activities

As in the previous example with Euler circuits, students first create their own algorithms and then they work with standard algorithms. They are not simply given algorithms to memorize and mimic. Situations involving minimal spanning trees provide a good opportunity for students to work with algorithms because several intuitive algorithms often arise naturally from students' exploration, and some of these algorithms work and some don't. Thus students are engaged in designing, using, and analyzing algorithms for solving interesting problems. This is what algorithmic problem solving is all about.

Since the nearest-neighbor algorithm does not work very well, students may wonder why they should study it. Teachers can cite several good reasons. First, the algorithm often comes up naturally in initial attempts to find minimal spanning trees (and in other graph-theory problems), and therefore it needs to be discussed and analyzed. Second, by comparing two different algorithms, students gain experience in the skill of algorithmic analysis. Third, it is important for students to understand that some algorithms may seem reasonable for a given problem but in fact may not work well. Finally, it provides an example of a situation that often occurs when algorithms are used to find optimal solutions: an algorithm may optimize locally, that is, at each step, and yet it may not optimize globally, that is, for the whole problem.

Algorithm efficiency is important to consider. Efficiency is a technical property that relates, roughly, to how fast an algorithm might run on a computer. If the run time grows no faster than some power of the size of the problem, then it is called a *polynomial-time algorithm.* The size of a problem is measured by, for example, how many edges are in the graph. Polynomial-time algorithms are deemed efficient, since they will terminate and produce a result for large problems in a reasonable time. If the run time grows exponentially with the size of the problem, or faster, then even if the algorithm works, it may take too long to produce the desired result. Kruskal's algorithm (the best-edge algorithm in activity 5) and Prim's algorithm (discussed above) are polynomial-time algorithms. The nearest-neighbor algorithm for minimal spanning trees does not work in general, so its efficiency is a moot point, although it too would fit in the polynomial-time category.

A technical analysis of efficiency is beyond the scope of most high school courses, but students should be alerted to the issue. They can begin to analyze efficiency technically when they investigate so-called brute-force methods for solving graph-theory problems. Using a brute-force method, you find and check all possibilities and then choose the best. However, there may be so many possibilities that to find and check them all could take centuries, even for the world's fastest computer. Thus a brute-force method will work only as long as a finite number of possibilities must be checked, and yet it may not be efficient. A brute-force method could be applied to the minimal-spanning-tree problem, but such methods are usually discussed in the context of problems that do not have known efficient solution procedures, like the famous Traveling Salesman problem. For an example of high school student activities related to brute-force methods and the Traveling Salesman problem, see Hirsch et al. (1998).

ALGORITHMIC PROBLEM SOLVING AND RECURSION

It is interesting to contrast the algorithmic problem solving at work in graph theory with the type of algorithmic problem solving that occurs in another area of discrete mathematics—recursion. Recursion is the method of describing sequential change by indicating how the next stage of a sequential process is determined from previous stages. Recursion naturally lends itself to an algorithmic analysis. However, whereas graph-theory algorithms are specific procedures that lead to solutions of particular problems like Euler-circuit algorithms or minimal-spanning-tree algorithms, recursive algorithms are of a more general nature, and in fact some graph-theory algorithms are recursive.

Treating recursion formally with technical symbolic representations (e.g., subscript notation and recurrence relations) can be very confusing to students. However, the concept of recursion can be succinctly captured in non-technical algorithmic terms. One way to do so for simple single-step recursion is to use the words NOW and NEXT, and make use of the built-in recursive capability of most modern calculators. Once students have a solid conceptual understanding of recursion, then the formal notation can be easily developed, as seen in the next example.

Sample situation

Wildlife management has become an increasingly important issue as modern civilization puts greater demands on wildlife habitat. As an example, consider a fishing pond that is stocked with trout from a nearby hatchery. Suppose that you are in charge of managing the trout population in the pond. Understanding how the population changes over time is important for effective management of the pond. How can you model and predict the changing trout population?

This situation can be viewed as involving continuous or discrete change. Taking the continuous viewpoint requires sophisticated tools from calculus and differential equations. Taking the discrete viewpoint, however, and considering this to be sequential change, say, year-to-year change, yields a much more accessible analysis and one that is often equally useful, if not more so.

The algorithms

The algorithms involved are not problem specific but rather a general type of algorithm, namely, recursive algorithms. For example, consider a simple population problem where initially 100 trout are in a pond and the population grows by an annual rate of 6 percent. The number of trout in the pond at the end of n years can be found in two ways. The so-called closed-form representation for the number of trout after n years is $A_n = 100(1.06)^n$. The recursive solution involves thinking about how the population grows from one year to the next. If NEXT is the number of trout next year and NOW is the number of trout this year, then an equation that models the year-to-year change is NEXT = 1.06NOW. In subscript notation, $A_{n+1} = 1.06A_n$, with $A_0 = 100$.

Using the closed-form representation, we can find the number of trout after, say, ten years by letting $n = 10$ in $A_n = 100(1.06)^n$. Using the recursive representation, we start with $A_0 = 100$, then multiply by 1.06 to get A_1, then multiply that answer by 1.06 to get A_2, and so on, until we arrive at A_{10}. This latter solution is a recursive (or iterative) algorithm. That is, it is a step-by-step systematic procedure in which the next step is determined by the previous step.

In the past, closed-form solutions have typically received greater attention, partly because of the cumbersome repeated computation required for the recursive solution. However, the computation is no longer cumbersome when technology is brought to bear. Moreover, the recursive formulation more clearly captures the essence of the process of sequential change. Furthermore, in many situations it is difficult or impossible to find the closed-form representation.

Some student activities

The following sequence of student activities illustrates how recursive algorithmic problem solving can be used to analyze the trout-population problem.

1. So far you have very little information about the fishing pond that you are supposed to manage. What additional information do you need in order to predict changes in the size of the trout population over time? Make a list.
2. A typical first step in mathematical modeling is simplifying the problem and deciding on some reasonable assumptions. Three pieces of information that you may have listed in activity 1 are (1) the initial size of the trout population in the pond, (2) the annual growth rate of the population, and

(3) the annual restocking amount, that is, the number of trout that are added to the pond each year. For the rest of these activities, use just the following three assumptions:

- There are 3000 trout currently in the pond.
- Regardless of restocking, the population decreases by 20 percent each year because of the combined effect of all causes, including natural births and deaths and trout being caught.
- At the end of each year, 1000 trout are added.

(*a*) Using the three assumptions above, determine the trout population in the pond after 1 year. After 2 years.

(*b*) What is the population after 3 years? Explain how you figured it out.

(*c*) Write an equation using the words NOW and NEXT that models this situation.

(*d*) Use the equation from step 2c and the last-answer key on your calculator to find the population after 7 years. Explain how the keystrokes you use on the calculator correspond to the words NOW and NEXT in the equation.

3. Think about the long-term population of trout in the pond.

   (*a*) Do you think the population will grow without bound? Level off? Die out? Make a conjecture about the long-term population. Compare your conjecture to those made by other students in your class.

   (*b*) Determine the long-term population. Was your conjecture correct? Explain, in terms of the fishing pond's ecology, why the long-term population you have determined is reasonable. (The long-term population levels off at 5000.)

   (*c*) Does the trout population change faster around year 5 or year 25? How can you tell?

4. A compact notation can be used for equations involving NOW and NEXT that allows for a more detailed analysis of the equations and the situations they model. In the context of the changing trout population, the subscript notation P_n can be used to represent the population after n years. (The notation P_n is read "P sub n.") Thus, P_0 is the population after 0 years, in other words, the initial population. P_1 is the population after 1 year, P_2 is the population after 2 years, and so on. Find P_0, P_1, P_2, and P_3 for the trout-population problem you have been studying.

5. This subscript notation relates closely to the way you have used the words NOW and NEXT to describe sequential change in different contexts.

   (*a*) In the context of a changing population, if P_1 is the population NOW, what subscript notation represents the population NEXT year?

   (*b*) If P_{24} is the population NEXT year, what subscript notation represents the population NOW?

(*c*) If P_n is the population NEXT year, what subscript notation represents the population NOW?

(*d*) The equation you found in activity 2 that models annual change in the trout population is NEXT = 0.8NOW + 1000. Rewrite this equation using P_n and P_{n-1}. (Note that students could equally well be asked to use P_n and P_{n+1}. P_{n-1} is used here to match the notation on certain graphing calculators.)

6. The equations in activity 5*d* do not tell you what the population is for any given year; they show only how the population changes from year to year. You can analyze situations like this that involve sequential change by using the "recursion mode" on your calculator.

   (*a*) Set the mode on your calculator so that you get equations that look like P_n=... or U_n=... or U(n)=.... Different calculators may have different names for this mode and the letters and notation may be slightly different, but they all allow you to use recursion.

   (*b*) Suppose U_n=... is the notation on your calculator. If U_n represents the population NEXT year, what subscript notation represents the population NOW?

   (*c*) Write the population equation NEXT = 0.8NOW + 1000 in the U_n=... form, and enter it into your calculator.

   (*d*) You can now use your calculator to produce a table and graph of n versus U_n. You do so by using the same keys and menus that you have used to produce graphs and tables for equations in the y=... form, except that you now have different settings to decide on. Make all the appropriate settings and produce a graph and table of n versus U_n.

7. Describe any patterns of change you see in the table and graph. Be sure to describe how the long-term population trend shows up in the table and graph.

Reflecting on the student activities

In these activities students use recursion to analyze sequential change (a changing trout population). They first use the words NOW and NEXT to formulate an equation that models the situation. This equation is in effect a recursive algorithm, which can easily be implemented on most modern calculators or on a spreadsheet. Students then formalize the statement of the algorithm (the NOW-NEXT equation) using subscripts, which allows the algorithm to be implemented more automatically by using the recursive mode of a calculator.

The NOW-NEXT formulation is an intuitive representation that makes the important concept of recursion easily accessible to students. Formal subscript notation is powerful but dangerous. If subscripts are introduced too soon, students may get bogged down in the technical formalism and miss the basic idea of recursion. If introduced after conceptual understanding is established, then the subscript notation can be very useful, both as a means for implementing the recursive algorithm and as a way to unify different topics. For example, the recursion equation used in the trout-population problem is an example of what is called an *affine recurrence relation*, $A_n = rA_{n-1} + b$. Varying the parameters r and b and considering closed-form and recursive formulations allows this single equation to refer to geometric sequences (let $b = 0$), exponential functions (closed-form formulas for the nth term of a geometric sequence), arithmetic sequences (let $r = 1$), linear functions (closed-form formulas for the nth term of an arithmetic sequence), and numerous applications, including compound interest, population growth, chemical concentration, and language learning.

The closed-form solution for the affine recurrence relation $A_n = rA_{n-1} + b$ is

$$A_n = \left(A_0 - \frac{b}{1-r} \right) r^n + \frac{b}{1-r}.$$

So, for the trout-population problem we get $A_n = -2000(0.8)^n + 5000$. Students can develop these closed-form solutions later, but we will surely overwhelm students if we introduce them too early. In any event, as students model and analyze problems by formulating and analyzing NOW-NEXT equations and the more formal recurrence relations, they are engaged in recursive algorithmic problem solving.

CONCLUSION

Algorithmic problem solving—the design, use, and analysis of algorithms to solve problems—is a powerful mode of problem solving. It is a particularly useful style of problem solving in discrete mathematics. Developing skill in algorithmic problem solving will contribute significantly to students' mathematical power.

REFERENCES

Dossey, John. *Discrete Mathematics and the Secondary Mathematics Curriculum.* Reston, Va.: National Council of Teachers of Mathematics, 1990.

Hart, Eric W. "Discrete Mathematics: An Exciting and Necessary Addition to the Secondary School Curriculum." In *Discrete Mathematicsa across the Curriculum, K–12,* 1991 Yearbook of the National Council of Teachers of Mathematics, edited by Margaret J. Kenney, pp. 67–77. Reston, Va.: National Council of Teachers of Mathematics, 1991.

Hirsch, Christian, Arthur Coxford, James Fey, Harold Schoen, Gail Burrill, Eric Hart, and Ann Watkins. *Contemporary Mathematics in Context: A Unified Approach, Course 2.* Chicago: Everyday Learning Corp., 1998.

National Council of Teachers of Mathematics. *Curriculum and Evaluation Standards for School Mathematics.* Reston, Va.: National Council of Teachers of Mathematics, 1989.

———. *Discrete Mathematics across the Curriculum, K–12.* 1991 Yearbook of the National Council of Teachers of Mathematics, edited by Margaret J. Kenney. Reston, Va.: National Council of Teachers of Mathematics, 1991.

Rosenstein, Joseph G. "Discrete Mathematics in the Schools: An Opportunity to Revitalize School Mathematics." In *Discrete Mathematics in the Schools,* edited by Joseph G. Rosenstein, Deborah S. Franzblau, and Fred S. Roberts, pp. xxiii–xxx. Providence, R.I.: American Mathematical Society, 1997.

ADDITIONAL READING

Consortium for Mathematics and Its Applications (COMAP). *For All Practical Purposes: Introduction to Contemporary Mathematics.* 4th ed. New York: W. H. Freeman & Co., 1997.

Core-Plus Mathematics Project. *Contemporary Mathematics in Context: A Unified Approach, Courses 1–4.* Chicago: Everyday Learning Corp. 1998–2000.

Sandefur, James T. *Discrete Dynamical Modeling.* New York: Oxford University Press, 1993.

Wilson, Robin J., and John J. Watkins. *Graphs: An Introductory Approach.* New York: John Wiley & Sons, 1990.

31

The Traveling Salesperson
Some Algorithms Are Different

Lowell Leake

MANY of us who teach or learn mathematics react to the word *algorithm* with memories of, for example, doing long division, finding square roots, bisecting an angle, and doing certain tasks in computer programming. This paper describes some algorithms that are totally different; they are also quite interesting, and they are definitely practical and widely used in important real-life applications. Best of all, they are easy to learn and understand.

Consider the district sales manager who is responsible for ten of the counties in the state. Workshops for training new personnel to introduce a new product have been scheduled in one city in each of the ten counties, and the manager has decided to start from home (in one of the ten cities), then drive to each of the other nine cities to set up the workshops, and then return home by the shortest route possible—an optimal solution. The route the manager seeks is known as a Hamiltonian circuit (i.e., it starts and ends at the same place and visits all the other places exactly once). This problem is an example of the famous Traveling Salesperson problem (TSP) from graph theory (Consortium for Mathematics and Its Applications 1997; Ore 1990; Tannebaum and Arnold 1992). Finding an optimal route is a very important application in many settings, especially in the business world. *Optimal* could mean shortest, quickest, cheapest, or the like. Other parts of graph theory involve optimization problems that deal with establishing telephone circuits between communities, finding a route that goes over each road exactly once and returns home (an Euler circuit), or producing printed electric circuits.

The manager's travel budget is limited, so she decided to plan the route so that the total mileage traveled by car is minimized. The manager noted that direct roads connected every pair of cities she needed to visit, which confirms that a Hamiltonian circuit exists. The manager thought, "This may be easy; I'll just list all the possible routes and pick the one with the least total mileage." This is the "brute force" method. Before the manager started the list, some combinatorial mathematics showed that there are 181 440 different possible routes for the ten cities. The formula $(n-1)!/2$ gives the number of routes for any set of n cities (Consortium for Mathematics and Its Application 1997; Ore 1990; Tannebaum and Arnold 1992). Division by 2 is required to eliminate routes that are merely reversals of direction of other routes and as such are not considered different.

The manager quickly realized that listing all routes for such a situation is totally impractical. With just twenty-five cities and a computer that can list 1 000 000 routes per second, for example, it would take ten billion years to generate all the approximately 3×10^{23} different routes (Consortium for Mathematics and Its Applications 1997). Algorithms need to be invented that will do the job very quickly, even with paper and pencil, and that will end up with a route that may not be the shortest one but that will be "close enough" for a practical solution. These algorithms are sometimes known as *greedy*, or *heuristic*. The manager knew that many mathematical algorithms give approximate answers instead of exact answers; for example, the long division algorithm rarely gives an exact answer for two numbers chosen at random and any square root algorithm gives an approximate answer unless the process is applied to a perfect square. The result of an arithmetic problem done on a calculator is often expressed as an approximation. Granted, those algorithms will find an answer that is as close as we want, whereas the algorithms for the TSP can be guaranteed only to find an answer that is "close enough for all practical purposes."

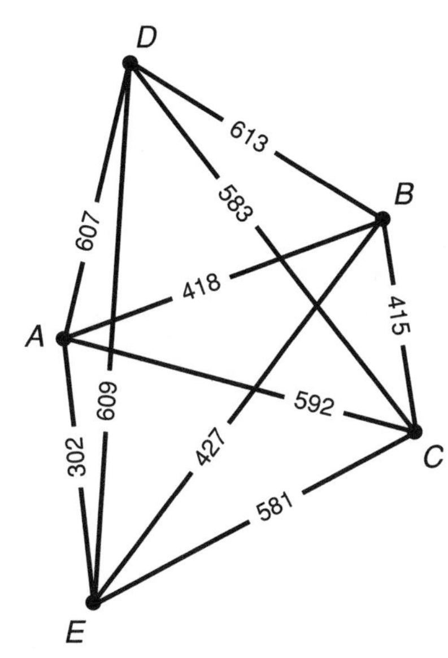

Fig. 31.1. A representation of the Traveling Salesperson problem involving five cities

The manager thought that a common-sense approach would be useful, so she devised the simpler situation, shown in figure 31.1, that involves only five cities. The cities are the vertices, and the roads are the edges. The reader should pretend that home base is *A* and use the formula above to discover that twelve different routes are possible: $(5-1)!/2 = 12$. The mileages are called *weights*. This term can also be applied to the cost of airline tickets or any other appropriate value. The manager thought of the following possible algorithm: "I'll just go to the city closest to *A* first, then go to the city closest to that city, and keep that up until I've visited all the other cities and must return home." Figure 31.2 demonstrates such a step-by-step solution with wavy lines. Starting at *A*, go first to *E*, then *B*, then *C*, then *D*, and finally back to *A*. This route is called *AEBCDA*, and the reverse, *ADCBEA*, would give the same mileage. For this reason, the formula divides $(n-1)!$ by 2. The mileage is 2334, but there is no way of knowing if 2334 is the minimum mileage unless all twelve possible routes are listed and all the mileages compared. In fact, of the twelve possible routes, the shortest mileage is 2327 and the longest is 2820, so this algorithm came within 7 miles of the optimal solution (not bad!), and it was found very quickly.

This technique of going to the closest city not yet visited and eventually back to home is known as the *nearest neighbor algorithm*. Students often discover the nearest neighbor algorithm on their own when asked to figure out a way to find the shortest route. Occasionally, a student will say, "Go to the farthest city first." This is a logical possibility, but it usually leads to poor results.

The completed diagram in the last step in figure 31.2 shows that all five cities in the example are on the route, which means that any traveling

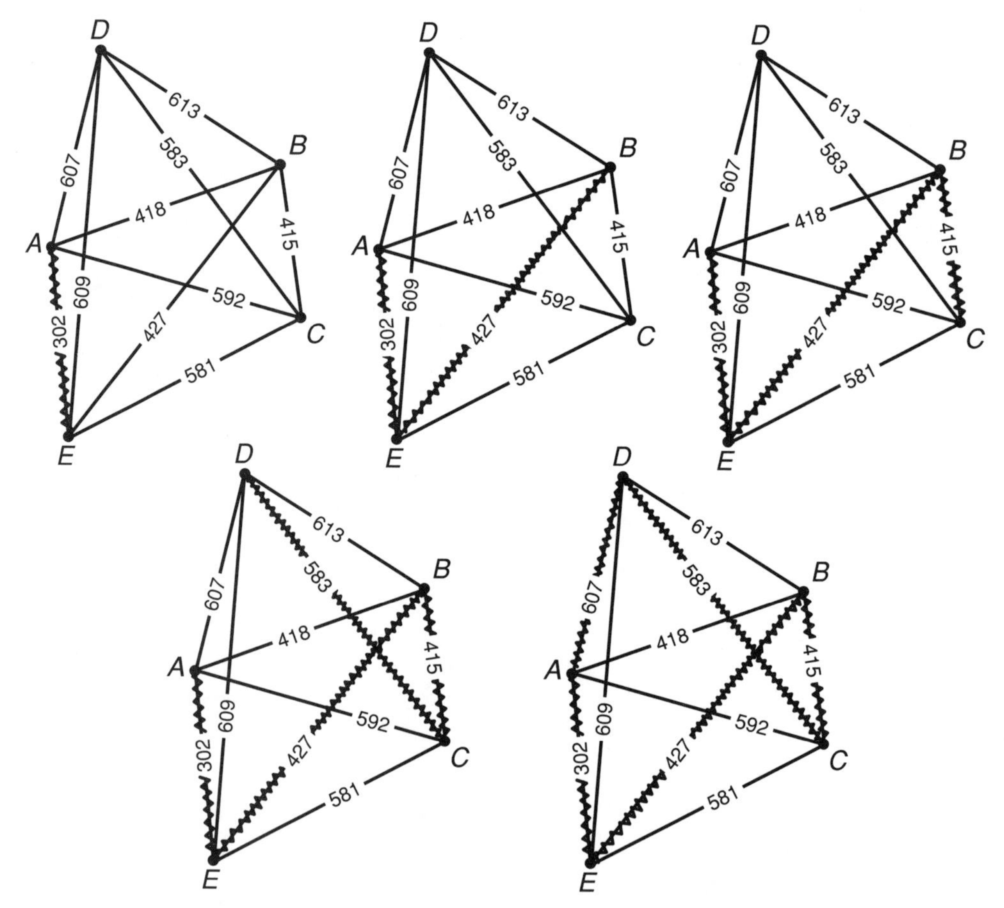

Fig. 31.2. A solution to the Traveling Salesperson problem produced by the nearest neighbor algorithm

salesperson could use that route regardless of where he or she lived. Readers will realize that they could use the nearest neighbor algorithm for each one of the five home cities and pick out the shortest of the five alternatives. This method is the *repeated* (or *repetitive*) *nearest neighbor algorithm.* Many teachers and students discover this variation on their own. In the example involving ten cities, the manager would just apply the nearest neighbor algorithm ten times and pick the shortest of the ten routes. In our five-city example, it turns out that the five answers are not all different—not an unusual outcome. Starting at *B*, the route found is *BCEADB*, whose mileage is 2518—a different route and different mileage from those for home base *A*. In the five-city example, the remaining three routes from starting points *C*, *D*, and *E* are *CBAEDC*, *DCBAED*, and *EABCDE*, respectively. These all have the same mileage (2327), which suggests they are identical except for the starting point. To see that they are identical, place the letters for starting point *C* (*CBAEDC*) in order on a circle. Next, start at *D* or *E*, go around the circle—perhaps in the other direction—and you have the same sequence. Therefore, for this example, the repeated nearest neighbor algorithm yields only three of the possible twelve routes. For thirteen cities, the algorithm would be repeated thirteen times, which is much more efficient than looking at all 239 500 800 possibilities. Will the thirteen routes all be different? It is not certain; the greater the number of cities, the higher the probability that the repeated method will give different results. Will one of the thirteen routes be optimal? Probably not, but we will never know without listing all the routes. However, a very reasonable short route will be found this way, and it can be found very quickly—an important factor in practical applications.

Another algorithm often used to solve the TSP is the *sorted edges algorithm.* It is not as intuitive as the nearest neighbor method, but it is just as easy. First, list the weights (mileages) in ascending order; that is, sort the edges. In the five-city example this order is 302, 415, 418, 427, 581, 583, 592, 607, 609, 613. The next step is to join together the two cities with the shortest distance (302) between them, *A* and *E*. Then join together the two cities with the second shortest distance (415), *B* and *C*. Continue until a route is completed. As in this problem, the first segments you draw may not be connected. Be careful to avoid the following two situations:

1. Three edges of the route come into one city. If this happens, it implies that the city is being visited more than once, a violation of the requirements for a TSP. If this situation occurs, skip the third link into the city and go on to the next longer distance.
2. Not all cities are included in a circuit. If this happens, a city has been visited twice. If a link produces a second visit, skip it and go on to the next longer distance.

The step-by-step sorted edges solution for the five-cities example is shown in figure 31.3. The solution is *ABCDEA*. The mileage—2327—happens to be optimal, but an optimal route can never be guaranteed. A thoughtful examination reveals that the sorted edges algorithm yields only one possible route, no matter where the salesperson lives. Many students are confused by this result. To avoid this confusion, it helps to break the class into five groups, one group for each hometown city, and let each group apply the algorithm. The students quickly catch on that it doesn't matter where the salesperson lives, each group does exactly the same thing, and there is only one answer. The students are also sometimes confused because as the solution takes shape, all the various links may not be connected until the final steps or because the evolving route may not include their hometown until the last step or two.

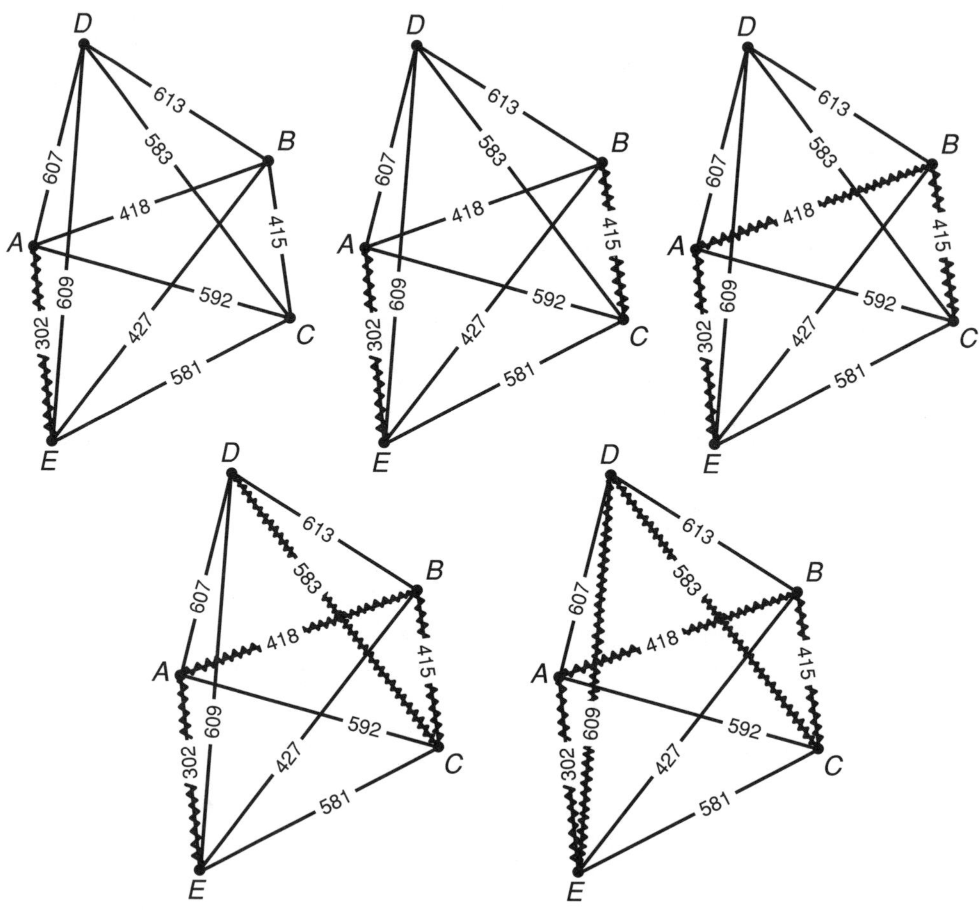

Fig. 31.3. The solution to the Traveling Salesperson problem produced by the sorted edges algorithm

Will a solution found with sorted edges be shorter than any nearest neighbor route? The only way to be certain is to compare it with all the repeated nearest neighbor solutions. In the five-city example, the optimal solution is equivalent to the three solutions (really only one) found by starting at *C, D,* or *E*. The sorted edges algorithm produces an optimal solution in this example, but only by chance. The greater the number of cities, the more remote the possibility of finding the optimal solution, and besides, we cannot be certain of having the optimal solution unless we list and compare all the solutions.

A wonderful source for examples is a standard road atlas that shows distances between cities on maps or in tables. Pick some cities that include the town where your school is located, or a nearby city. Give distances in either miles or kilometers.

Finally, here are two alternative algorithms the reader can try:

1. At each step of the nearest neighbor algorithm, go to the city closest to either end of the route determined at that stage. Try this starting at *B* in figure 31.1.
2. First link the two closest cities, as in the sorted edges algorithm. Then continue in the manner described in the previous algorithm, using the nearest neighbor approach.

REFERENCES

Consortium for Mathematics and Its Applications. *For All Practical Purposes.* 4th ed. New York: W. H. Freeman & Co., 1997.

Ore, Oystein. *Graphs and Their Uses.* Revised and updated by Robin J. Wilson. Washington, D.C.: Mathematical Association of America, 1990.

Tannebaum, Peter, and Robert Arnold. *Excursions in Mathematics.* Englewood Cliffs, N.J.: Prentice Hall Press, 1992.

32

Using Algorithms to Generate Objects of Mathematical Interest

Elaine Simmt

DAVIS and Hersh, in their book *The Mathematical Experience* (1981), contrast algorithmic mathematics with dialectic mathematics: algorithmic mathematics, they say, generally provides results of some sort, whereas dialectic mathematics generally leads to establishing whether or not a solution exists. They suggest that these two forms of mathematics are based on different approaches and responses to mathematical problems. "*Dialectic* mathematics invites contemplation. Algorithmic [sic] mathematics invites action. *Dialectic* mathematics generates insight. *Algorithmic* mathematics generates results" (emphasis in original) (Henrici [1974] cited in Davis and Hersh [1981, p. 183]). They claim that both paradigms have existed in mathematics over the last century but until recently, the dialectic paradigm has been more prevalent than the algorithmic.

In school mathematics, learning algorithms has been a significant component of the curriculum for years. For the most part, however, algorithms have been taught simply as ways of finding solutions to mathematics questions. Consider, for example, long division, synthetic division, or Gaussian elimination (Hart 1991; Maurer and Ralston 1991). Unfortunately, students often use algorithms rotely to find answers to questions instead of meaningful solutions to problems.

The use of algorithms that I am recommending for high school mathematics is found in the work of the research mathematics community—in particular, the work of fractal geometers. Rather than use algorithms simply to determine a result, these mathematicians use computer algorithms to generate objects, such as the Mandelbrot set, that are of mathematical interest. The important point here is that the result produced by the algorithm is *not an endpoint* of the fractal geometer's work; rather, it is *a starting point* from which the geometer engages in mathematical explorations and dialectical thinking.

For instance, consider the algorithm for the famous Koch curve. Begin with a line segment and on the middle third of it place an equilateral triangle; then remove its base. Repeat this process on the remaining line segments (see fig. 32.1). This algorithm generates an object with a very interesting property. Although the curve is continuous, it cannot be differentiated. This property led the mathematics community to view it as a monster. Today, there is a whole class of such curves, and they are considered to be very important for the study of fractals and dynamic systems. This is a simple example; however, it demonstrates how using algorithms differently can lead to understanding differently. The distinction between algorithmic and dialectical thinking becomes blurred when the algorithmic thinking is of the kind noted here.

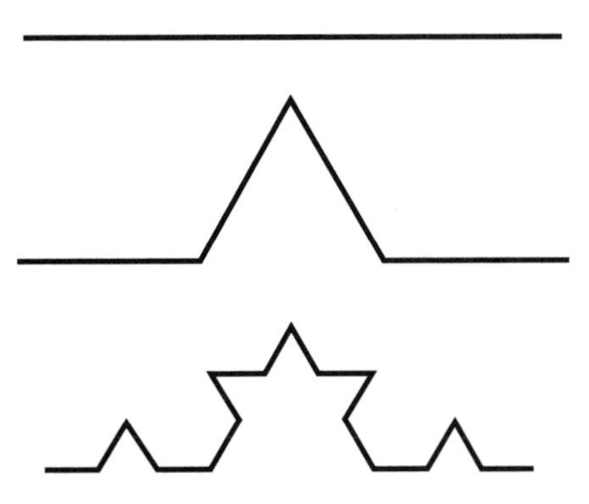

Fig. 32.1. The initial segment and first two stages of the Koch curve

USING ALGORITHMS DIFFERENTLY

Using algorithms as a means of generating objects to be investigated has the potential to make a number of topics taught in secondary school mathematics more accessible and more meaningful to students. The following activities demonstrate how algorithms can be used to generate mathematical spaces for dialectical thinking. I will discuss one activity in detail and then note some other examples of using algorithms in this way (see Peitgen, Jurgens, and Saupe [1992] for further examples). Each of the activities has two parts: the first involves the construction, by means of an algorithm, of a fractal or fractal-like object, and the second involves an investigation of that object. These activities are suitable for high school students, but they could also be used with younger students by simply posing different questions.

Fractal Cards

This activity is based on fractal-like paper constructions originally developed by the industrial designer Uribe (1995). By applying a simple set of procedures involving cuts and folds to a sheet of paper, students construct a fascinating pop-up card on which they can base an investigation of concepts from areas such as geometry, measurement, and discrete mathematics. The physical actions of cutting and folding and the recursive algorithmic nature of this process provide an opportunity for students to think differently about relationships and patterns that in the past they have explored with paper and pencil.

To construct the card (see fig. 32.2), use a simple algorithm:

1. Begin with a sheet of paper folded in half.
2. Along the fold make two perpendicular cuts one-quarter of the way in from each side and one-half the way up the sheet.
3. Fold to the edge of the cuts.
4. Repeat the cut-and-fold process on the newly formed fold.
5. Repeat the process until the paper can no longer be cut because of its thickness.

Once no further cuts can be made, the folds must be unfolded and the "fractal" pushed out to form a relief image when the paper is opened to 90 degrees.

When repeated, the simple rule (fold and make two cuts) generates a fractal limit curve in the form of a set of stacked cells (see fig. 32.3). By following the algorithm that involves iterating this simple rule, a compelling object called a *fractal card* is created. The fractal card is not a product in the sense that some problem is solved or some question answered; rather, the fractal card becomes the subject for students' mathematical thinking.

Make two cuts along the fold of the paper 1/4 of the way in from each side and 1/2 the way up the sheet.

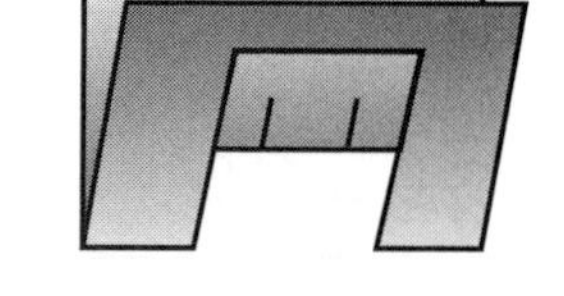

Fold the paper up to the cuts, then repeat the cut-and-fold rule.

Fig. 32.2. An algorithm for constructing a fractal card

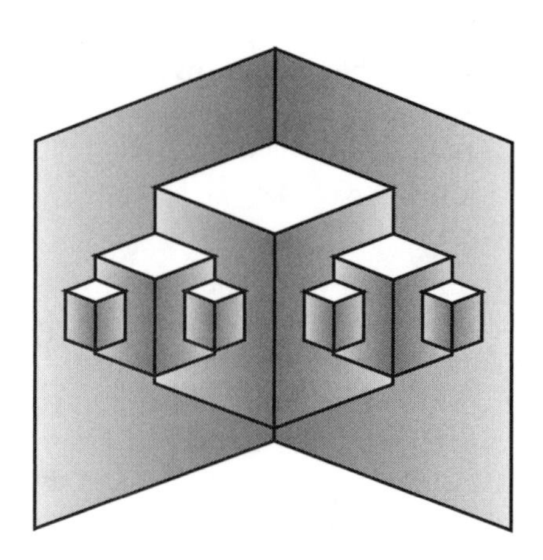

Fig. 32.3. A completed fractal card

Students might begin an investigation of the mathematics of the fractal card by considering the growth of the cells. They could generate a table of values by counting the cells at each stage and then generalize the growth pattern. They should note, for the given card, that the number of cells generated at stage n is 2^{n-1} and the total number of cells is found by computing $2^n - 1$. A related question is to determine the total number of cuts. If students flatten the card, the cuts become easy to count, and again a generalization can be made. The number of cuts at stage n is given by 2^{n+1}, and the total number of cuts is simply $2^{n+1} - 2$. By assigning a value (let's say 2) for the area of the largest cell, students can determine the formula for the total surface area of the cells by using

$$SA_n = 2\sum_{i=1}^{n}\left(\frac{1}{4}\right)^{i-1},$$

where i represents the stages from 1 to n and n represents the total number of stages, and then determine the behavior of the surface area as the number of cells approaches infinity. This particular question offers a wonderful occasion for dialectical thinking, since students could deduce that the surface area of the cells must be finite (converge) because the area of the paper is finite. Conversely, students are often surprised to discover that the total length of the cuts diverges as n approaches infinity. A finite area contained by an infinite perimeter seems paradoxical; yet it is clear that this is so, since the total length of the cuts is given by the divergent series

$$1 + 2\left(\frac{1}{2}\right) + 4\left(\frac{1}{4}\right) + 8\left(\frac{1}{8}\right) + \cdots + n\left(\frac{1}{n}\right).$$

Students could be encouraged to think of other questions by simply having them tip the card so that they see a cross section of the cells. Then they might consider the quadrilaterals that are formed by the plane that intersects the cells at right angles. Students can play with the card by opening and closing it. They might note that the areas of the quadrilaterals on this plane are minimized when the two sides of the card are pulled down or when the card is closed and that the maximum area of the quadrilaterals occurs when the card is opened at 90 degrees, an action that creates squares. These are just a few examples of the kinds of mathematical concepts that can be addressed in the activity described here. By creating new algorithms for the cuts and folds, students can generate other fractal cards and find different patterns and concepts to explore. (See Simmt and Davis [1998] for related activities.)

The Cantor Set

In this exercise, students use a simple paper-and-pencil algorithm to produce an object of considerable mathematical interest. Draw a line segment to represent the interval $0 \le x \le 1$. Next, erase (remove) the middle third, $1/3 < x < 2/3$, of that line segment, which leaves the endpoints. Repeat this algorithm on the remaining line segments ad infinitum (see fig. 32.4). The end result is known as the *Cantor set,* or *Cantor dust.* Even junior high school students can consider some very interesting questions about the set and generating the set. For example, they can determine that the number of intervals generated at each iteration, I_n, is 2^{n-1}. Students could also estimate, with a ruler and then numerically, the total length of the missing interval (erased space) at each iteration: $L_n = L_{n-1} + (1/3)^n \cdot 2^{n-1}$. In addition to considering these questions, older students could be challenged to think about what is left in the set as the number of iterations approaches infinity and to identify points remaining in the dust, such as the endpoints of each interval. In this exercise we see how the alternative use of an algorithm creates an opportunity for students to think about powerful ideas in mathematics—in this example, a set that is almost all holes yet is so large that it is uncountable.

Fig. 32.4. The Cantor set

The Koch Snowflake Curve

One of the most famous of fractals is the Koch snowflake curve (see fig. 32.5). To construct this fractal, begin by drawing an equilateral triangle (interesting distortions are made when you begin with any other triangle). Replace the middle third of each side of the triangle with two segments of the same length, each rising 60 degrees from the original segment, to form a peak on the side of the triangle. Each iteration consists of repeating this procedure on the newly formed sides of the emerging snowflake. Students can be asked to determine the number of sides at any stage of the snowflake. Thus $S_n = 3 \times 4^n$. They might also consider the perimeter. If each side of the original triangle has length 1, then the side lengths are, progressively,

$$1, \frac{1}{3}, \frac{1}{9}, \frac{1}{27}, \ldots,$$

or

$$\frac{1}{3^0}, \frac{1}{3^1}, \frac{1}{3^2}, \frac{1}{3^3}, \ldots.$$

Then the perimeter, P^n, can be given as

$$S_n \times \frac{1}{3^n},$$

or

$$3 \times 4^n \times \frac{1}{3^n}.$$

Once again, students' intuitions are challenged when they compare the perimeter and the area of this object. As the number of iterations approaches infinity, the perimeter of the Koch snowflake diverges but the area converges.

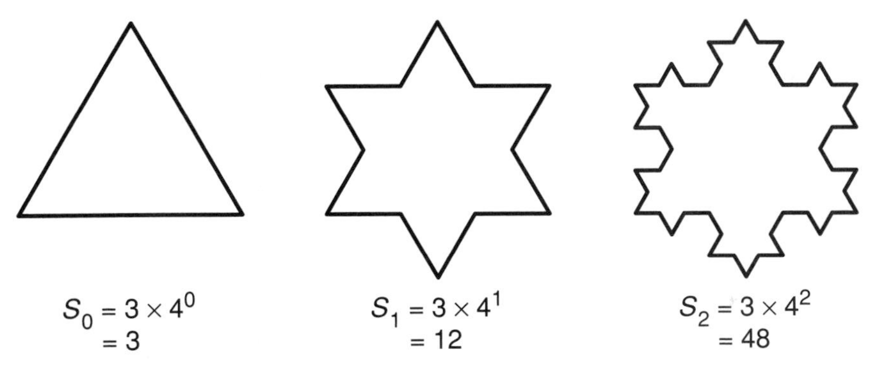

Fig. 32.5. These iterations of the Koch snowflake demonstrate the growth in the number of sides in the fractal

WHY USE ALGORITHMS IN THIS WAY?

The purpose of this paper has not been to suggest that we no longer use algorithms in traditional ways—to generate solutions, to program computers, to carry out processes, or to understand processes. Rather, I have suggested another use of algorithms found in the research-mathematics community but not commonly seen in school mathematics. This alternative use of algorithms is distinguished from traditional uses in that the product of the algorithm becomes the object of mathematical investigation. That is, the "mathematical" part of the activity is not in carrying out the algorithm; rather, it is in exploring the mathematics of the product of the algorithm.

In my experience, students who have participated in the activities described here are quite taken not only by the beauty of the fractals and the surprising results but also by the variety of mathematics found in the activities and by the new understandings they gain of some of the mathematics. For instance, in the fractal-card activity, the actions of cutting and folding provide students with an embodied meaning for a growth series. Students are not surprised to find a growth pattern in the cells when the card is opened because their physical actions gave them insight into this feature as they constructed the card. Further, having a physical experience and a concrete object (manipulative) on which to base discussions of topics in high school mathematics (e.g., sequences and series) facilitates a different kind of discussion among mathematics students than one that is based on reading printed text on a book page. Although we tend to provide embodied experiences for younger students' mathematics learning, we must attempt to seek these kinds of experiences for older students as well.

Using algorithms in the ways described also affords the opportunity for facilitating practice, making connections across the mathematics curriculum, and introducing new mathematics to students. Recording the sequences found in the growth patterns, generating the formulas for the series, and determining end behaviors engage students in meaningful practice and practices. One of the most compelling features of these activities is an abundance of opportunities to make connections between concepts and topics in mathematics. For example, students may make connections among measurement, sequences and series, geometry, and limits. Finally, these activities, and specifically this particular use of algorithms, provide students with an opportunity to think about fractal geometry and iteration. Not only are these two topics significant because they are found in present-day research mathematics, but they are very accessible and compelling to secondary school students.

Activities such as the ones described in this paper seem to motivate students and enhance their mathematical understanding. The reason may be because both dialectical and algorithmic thinking are encouraged; it may be because the mathematics is embodied in physical activity. Whatever the reason, using algorithms differently in school mathematics has the potential to occasion meaningful mathematical experiences for students.

REFERENCES

Davis, Philip J., and Reuben Hersh. *The Mathematical Experience.* London: Penguin Books, 1981.

Hart, Eric W. "Discrete Mathematics: An Exciting and Necessary Addition to the Secondary School Curriculum." In *Discrete Mathematicsas across the Curriculum, K–12,* 1991 Yearbook of the National Council of Teachers of Mathematics, edited by Margaret J. Kenney, pp. 67–77. Reston, Va.: National Council of Teachers of Mathematics, 1991.

Henrici, Peter. "Computational Complex Analysis." In *The Influence of Computing on Mathematical Research and Education,* edited by J. P. LaSalle, pp. 79–91. Proceedings of a Symposium in Applied Mathematics, vol. 20. Providence, R.I.: American Mathematical Society, 1974.

Maurer, Stephen B., and Anthony Ralston. "Algorithms: You Cannot Do Discrete Mathematics without Them." In *Discrete Mathematics across the Curriculum, K–12,* 1991 Yearbook of the National Council of Teachers of Mathematics, edited by Margaret J. Kenney, pp. 195–206. Reston, Va.: National Council of Teachers of Mathematics, 1991.

Peitgen, Heinz-Otto, Hartmut Jürgens, and Dietmar Saupe. *Fractals for the Classroom.* New York: Springer-Verlag, 1992.

Simmt, Elaine, and Brent Davis. "Fractal Cards: A Space for Exploration in Geometry and Discrete Mathematics." *Mathematics Teacher* 91 (February 1998): 102–8.

Uribe, Diego. *Fractal Cuts.* Norfolk, England: Tarquin Publishers, 1995.